LONGSHOT

LONGSHOT

The Inside Story of the Race for a COVID-19 Vaccine

DAVID HEATH

NASHVILLE NEW YORK

Center Street
Hachette Book Group
1290 Avenue of the Americas, New York, NY 10104
centerstreet.com
twitter.com/CenterStreet

First Edition: January 2022

Center Street is a division of Hachette Book Group, Inc.

Print book interior design by Timothy Shaner, NightandDayDesign.biz

Library of Congress Control Number: 2021948602

ISBNs: 9781546000907 (Hardcover); 9781549164682 (Audiobook); 9781546000921 (E-Book)

Printed in the United States of America

LSC-C

Printing 1, 2021

For Peter J. Connelly, who taught me to love writing.

CONTENTS

LONGSHOT

INTRODUCTION

The COVID-19 vaccine is one of the greatest achievements in modern medicine. For the first time in history, we created vaccines quick enough to tame a pandemic. The speed with which they were developed was, by historical standards, unparalleled. The polio vaccine, which has nearly eradicated a crippling disease that even afflicted a US president, was 50 years in the making. Despite herculean efforts over four decades, the creation of an effective HIV vaccine still eludes some of the most brilliant minds in medical research. Yet just days after China disclosed the true nature of a mysterious outbreak in Wuhan, Anthony Fauci, director of the National Institute of Allergy and Infectious Diseases, predicted in what was left of his childhood Brooklyn accent that we would have a COVID-19 vaccine in just 12 to 18 months.

Some of his peers considered the prediction a longshot. "When Dr. Fauci said 12 to 18 months, I thought that was ridiculously optimistic and I'm sure he did too," Paul Offit, one of the nation's premier vaccine experts, told CNN at the time. Merck CEO Kenneth C. Frazier, who tried in vain to make its own vaccine, said in July 2020, "I think when people tell the public that there's going to be a vaccine by the end of 2020, for example, I think they do a grave disservice to the public."[1] Michael

Osterholm, a preeminent public health expert at the University of Minnesota, said, "The goal of 18 months is one that will be very, very difficult to achieve. But it just may be our moon shot."[2]

And yet, to everyone's surprise, Fauci's prediction turned out to be conservative. It took exactly 338 days from the time the virus was identified to the day Sandra Lindsay, an ICU nurse at Long Island Jewish Medical Center, became the first American to get the shot outside of a clinical trial.[3] Those desperate to be immunized scrambled for scarce appointments for the first few months. Yet, over time, an astounding number of people chose not to get vaccinated despite the grave risks the virus poses. By October 2021, only 57% of the population had been fully vaccinated, and the fiercely contagious Delta variant brought back the pandemic with a vengeance. While the unprecedented speed of development was a miracle for many, for others, it was a cause for suspicion. People from all walks of life, even health-care workers, began asking aloud how the vaccines could have been made so quickly? Had corners been cut?

The truth is the science behind the vaccines had been in the works for at least 15 to 20 years. Both the Moderna and Pfizer's mRNA vaccines relied on a breakthrough that was published by scientists at the University of Pennsylvania in 2005. Work on a coronavirus vaccine really began in 2013 with the outbreak of a different novel coronavirus: Middle East respiratory syndrome—MERS, for short. In fact, the COVID-19 vaccines stem from a scientific article published in August 2017[4] as well as a patent a team of a dozen scientists filed for in October 2017.[5]

For those inclined to believe conspiracies, this may seem suspicious. No, the scientists did not have advanced knowledge of SARS-CoV-2. The vaccine that we rolled up our sleeves for was essentially a MERS vaccine that had been designed in animal studies years earlier. It just needed to be tweaked. Indeed, had the virus made the fateful leap from bats to humans 10 years earlier—or even just five—science would not have been ready. The vaccines could not have been developed seemingly at the speed of light. The shutdown, as awful as it was, would have gone on months—if not years—longer. Millions of additional lives would have been lost. New variants pose a real risk, and the Delta variant brought back masks and social distancing. But the vaccines have so far remained remarkably effective. The hope is that booster shots will guard us against future variants. Supply remains a problem in the developing world. But in the United States in particular, the real problem is not the vaccine; it is people's refusal to get vaccinated.

Much of the reporting on the vaccines has been focused on the achievements of Operation Warp Speed or the vaccine makers themselves. Their contributions were enormous, especially in compressing the timeline of development and paying up front for vaccines that might not work. However, the truth is that the politicians and the drug companies could not have made these vaccines so rapidly had it not been for a small group of scientists who made the right decisions at the right time and may ultimately save the world from an even more devastating pandemic. Their stories are ones of incredible foresight as well as incredible luck. Their breakthroughs had quietly revolutionized vaccine science, and yet hardly anyone realized it or even frankly cared about it at the time. Even the scientific community was largely unimpressed as it was happening. Prestigious scientific journals rejected some of the key articles that would ultimately make the vaccines possible. This book will tell the stories of those scientists rather than the vaccine makers, with one exception. Moderna was chosen by government scientists to be the platform for the first COVID-19 vaccine because it offered such a speedy way of making vaccines.

The company had been working closely for years with the National Institutes of Health on a coronavirus vaccine. So its story is integral to the story of the science behind the vaccines.

One man in particular, Barney Graham at the National Institutes of Health, pulled all the threads together. He is the Jonas Salk of COVID-19 vaccines. His research into a coronavirus vaccine started nearly seven years before this new strain would appear in Wuhan, China. His main motive for studying coronaviruses was less to save the world and more to explore the unknown. We knew so little about that type of virus, and that made the research more exciting, more likely to break new ground. But in addition to the potential for personal recognition, Graham also understood that it was just a matter of time before we would see a new strain of coronavirus. There had been two novel strains of coronavirus in just a decade, and both of those—SARS and MERS—were frighteningly deadly. Fortunately, neither was terribly contagious. Given that there are thousands of additional strains in bats, it was a matter of when—not if—there would be another leap from animals to humans. If a new strain appeared that was highly contagious, it could prove devastating. It was critical to be prepared for the next one.

The emergence of HIV in the early 1980s and the unprecedented effort to develop a vaccine also played a pivotal role when SARS-CoV-2 appeared. If you tried to develop a perfect virus to elude vaccines, you would come up with something like HIV. It has the rare ability to play hide-and-seek with antibodies. And even if the antibodies found the virus, HIV magically disguises itself through rapid mutations, rendering the latest batch of antibodies useless. Today, HIV is controlled with antiviral drugs that allow those carrying it to live normal lives. The drugs are so effective that those infected cannot transmit the virus and even sensitive tests cannot detect the virus in those taking the medication. But the virus quickly comes back if you stop taking the antiviral drugs. That's why the quest for a vaccine is still ongoing. It took every bit of scientific know-how to come up with experimental vaccines, which to date have failed. But the science that went into that effort has changed our ability to develop vaccines forever.

While HIV is a uniquely challenging virus, SARS-CoV-2 is a rather easy virus to defeat. With the proper antibodies ready to pounce, SARS-CoV-2 and many of its variants are quite helpless. Not only did the mRNA vaccines prevent nearly all hospitalization and death, but they prevented symptomatic disease more than 90% of the time until the arrival of Delta. And beyond that, real-world data showed that they even kept those who were vaccinated from being effective carriers of the virus, at least with the early variants. Even with the Delta variant, breakthrough cases remain the exception. While it is difficult to quantify, this was almost certainly due to the caliber of vaccine science today. Graham turned his attention to coronaviruses after spending more than a decade trying in vain to make an HIV vaccine. He understood viruses with a level of precision previously unimaginable. As Paul Duprex, director of the Center for Vaccine Research at the University of Pittsburgh, would later say, "In 2020 we can make vaccines in ways that Jonas Salk could only have dreamed of."[6]

The two scientists most responsible for making mRNA an effective approach for vaccines are Katalin Karikó and Drew Weissman at the University of Pennsylvania. Weissman had worked in Fauci's lab just as HIV first emerged among the gay community in New York and California. Fauci would become director of the National Institute of Allergy and Infectious Diseases and the face of public health at the National Institutes of

Health. Weissman decided to continue to research this dreadful virus when he moved into academia. That ultimately led him to collaborate with Karikó, who had devoted her career to trying to solve the riddle of how to use RNA to treat disease. They figured out how to use mRNA without our bodies rejecting it. And yet, Penn officials, not realizing the significance of this discovery, sold the rights to it to an obscure company for next to nothing. Karikó was even shoved out the door by her school after her key discovery. Today, Karikó and Weissman are often mentioned as potential Nobel Prize winners. But at the time, their work was overlooked and even dismissed. Moderna wouldn't exist without their work. Yet the company writes them out of its history, insisting it was Moderna that made the key scientific breakthroughs in RNA science.

The story of the vaccines is full of what-ifs. At several stages, any wrong turn might have taken us down a path that would have deprived us of a COVID-19 vaccine. Weissman and Karikó were never able to convince investors to let them commercialize their discovery. What if another scientist—Derrick Rossi of Harvard University—had not used Weissman and Karikó's science to found Moderna? What if Graham had kept pursuing an HIV vaccine and never bucked the culture at the National Institutes of Health by turning his attention to other viruses? What if MERS had not broken out in the Middle East just when Graham was looking for a new virus to conquer?

Perhaps the biggest stroke of luck was that SARS-CoV-2 was not more lethal. What if, like SARS-CoV-1, its case fatality rate was 15%? Even worse, what if it had the killer instincts of MERS, another novel coronavirus that emerged in 2012? That virus, which still circulates today, kills 35% of its victims.[7] Imagine 5 million to 12 million deaths in the United States. The worst case would be akin to everyone in New York City and Chicago dying. Or imagine 42 million to 98 million deaths worldwide.

The brutal truth is that this is not likely the last pandemic we will face. There is a high probability that we will see new strains of novel coronaviruses in the future. With luck, we will be able to ramp up new vaccines quickly to defeat those. But there is also the likelihood that we'll see other viruses causing pandemics. Perhaps the next one will be a novel influenza. Or perhaps a new strain of Nipah, a particularly dreadful virus that served as the basis of the Hollywood thriller *Contagion*, starring Kate Winslet and Matt Damon. There are lessons to be learned from the discovery of the COVID-19 vaccine that could lead us to rethink the resources we devote to stopping future pandemics. As you'll learn from this book, we were lucky this time.

CHAPTER 1

Ready for a Pandemic

A colleague shot an urgent email to Barney Graham at 10:53 p.m. on a Friday night. Graham had been expecting the news for days—10 days to be exact. It had been that long since the world first learned of a mysterious new virus causing severe cases of pneumonia in Wuhan, China. The news broke the day the world was waiting to ring in the New Year. Now, on January 10, 2020, Graham took note of the subject line: "Wuhan Spike sequence." He had been trying without luck to get this information from China's chief public health official, George Gao, through back channels. But now it was published for the whole world to see. Someone had drawn a blood sample from a patient in Wuhan and used it to unlock the virus's genetic code. A professor at Shanghai's Fudan University posted it on Virological.org, a scientific bulletin board.

Most of the email was a jumble of four letters, A, C, G, T. These are abbreviations for the four bases of the virus's genetic code. To the naked eye, this mishmash of letters was meaningless, even for the most learned scientist. But when analyzed on a computer, it revealed the virus's identity. The email was brief.

Vincent Munster, chief of a virology lab at National Institutes of Health, had one of his scientists run the sequence through a database of known viruses. "Initial analysis shows it is about 73% identical to SARS . . . Closest relative when blasting the whole genome is a SARS-like CoV called ZC45." In short, it was a novel strain of coronavirus. The closest match had been identified in February 2017 from a bat.[1]

Graham had already suspected it was a coronavirus. But he was more interested in another bit of data. The genetic code also revealed the virus's Achilles' heel—the precise makeup of the so-called spike protein on its surface. Block that protein and you can render the virus harmless.

Graham, who was deputy director of the National Institutes of Health's Vaccine Research Center, knew exactly what to do. After two previous novel coronavirus outbreaks—SARS in 2003 and MERS in 2012—Graham had spent years getting prepared for the next novel strain. He went to bed confident that, in the morning, he would whip up the genetic code for a vaccine that the untested company Moderna would manufacture. Graham had no idea, however, just what a profound impact his years of expertise were about to have on the planet. At that moment, there were only 41 confirmed cases in Wuhan, although the virus had spread to other cities. There was speculation that the virus originated from a seafood market in Wuhan that also sold exotic meat. Wild animals such as the king rat snake, the Chinese bamboo rat, and the marmot were kept alive in cages and killed in front of customers to ensure freshness. But the conditions raised serious questions about hygiene.[2] Some scientists speculated that the virus might not spread from human to human. But Graham did not believe it. He could not imagine so many cases already coming from one meat market, no matter the conditions.

Still, there was much that Graham could not have known. He had no idea that this new coronavirus would unleash a pandemic and certainly not one of the deadliest pandemics since 1918. Likewise, he did not know he was about to change the course of medical science and potentially save millions of lives in the process.

For years, Graham had been growing more confident he could create a vaccine fast enough to stop a pandemic. That is a feat many of his colleagues might have thought absurd. But in Graham's mind, he just needed the right virus to prove it. History was not on his side. Typically, it takes years and even decades to bring a successful vaccine to market. Graham's quest seemed akin to Charles Lindbergh dreaming of flying to Paris at supersonic speeds. But Graham had a unique vantage point. He sat at the command center at the Vaccine Research Center for combating emerging disease. In fact, that was his precise job.

The center's primary mission from the start had been to develop an HIV vaccine. But the virus that causes AIDS—once a death sentence—proved a formidable foe. Our bodies mount a pitiful defense against it. Any vaccine able to defeat it would have to dramatically outperform our own immune systems. For any randomized clinical trial, there is an independent data safety monitoring board. These are the only people who know which volunteers got the actual drug and which ones got the placebo. They can stop a trial if anything serious happens. And in April 2013, the National Institutes of Health stopped a four-year clinical trial with more than 2,500 volunteers testing the latest experimental HIV vaccine. It had not just failed; it had failed miserably. Volunteers getting the placebo were slightly more likely to get HIV.[3] The difference in outcomes may have only been by chance, but it was enough to shut down the experiment. This was not the first defeat in finding an HIV vaccine, but it was a turning point for the Vaccine Research Center.

The center's scientists had learned so much about infectious diseases, but their attempts to create an effective vaccine were stymied by a virus with the surreal resilience of Arnold Schwarzenegger in the role of the Terminator. In its 13 years of existence, the Vaccine Research Center had not produced a single vaccine. Graham himself was ready to turn his advanced knowledge of vaccines on other viruses. He had a career-long obsession with respiratory syncytial virus. RSV is known best as a leading cause of colds, especially in children. But it is also the leading cause of hospitalizations in young children, and it rivals influenza as a cause of death among the elderly.

Part of what made RSV so intriguing was that a vaccine trial in 1966 had backfired and tragically made children more susceptible, much more likely to be hospitalized and gasping for breath. It's a problem within the vaccine world known as vaccine-associated enhanced disease. Something about the vaccine made the virus more likely to invade cells—the opposite of what it was supposed to do. Graham decided to make solving the riddle of what went wrong his personal cause. He devoted decades to it, and by 2013, Graham was confident he and his team had found the answer, borrowing from the techniques being used in HIV research.

His legacy might have been as the guy who solved the riddle of RSV. But Graham realized that the lessons he had learned could be applied to other viruses as well. And he wanted to combine his knowledge of how to defeat a virus with new technology that made making a vaccine much speedier and more precise. That's what ultimately led him to choose messenger RNA as the vehicle for a vaccine to stop a pandemic.

By 2017, Graham was working with Moderna on his concept of pandemic preparedness. He used the prototype pathogen approach. What that means is that you can study any virus within a given family, such as coronavirus, and make a vaccine that can be quickly adapted when a new strain emerges. They focused on two very deadly viruses, testing vaccine candidates in mice. One was a coronavirus, the Middle Eastern respiratory syndrome, or MERS. It was limited mostly to Saudi Arabia and had a case fatality rate of 34%. As of June 2021, there have been a total of 2,574 cases of MERS reported, with 886 deaths.[4] The other was Nipah, another virus that leaps from bats to humans. A 2018 outbreak in Kerala State, India, killed 21 of the 23 people infected, causing panic in the local community.[5] Neither disease was highly contagious, but that could always change with a new strain.

By 2019, Graham was ready to start proving his point. He came up with a plan—in essence—to click a stopwatch to show how fast he could make a vaccine and get it into humans. He was able to scrouge up $5 million for this demonstration project. He figured he would be able to go from a virus's genetic code to a human trial in just 100 days. This first step in clinical trials tests the safety and tolerability of different doses of a vaccine and evaluates how well the antibodies generated neutralize the virus in a test tube.

The two most obvious candidates for this demonstration were MERS and Nipah, the viruses Graham already understood so well. Graham chose the latter, and he was on the precipice of developing an experimental Nipah vaccine. But he hadn't fully decided on Moderna as the collaborator for this experiment. After all, Moderna had never produced an FDA-approved vaccine. In November 2019, Graham visited Moderna's new 300,000-square-foot manufacturing facility in Norwood, Massachusetts, and was impressed. He attended a meeting with Moderna CEO Stéphane Bancel the following month in Building 31 at the northern tip of NIH's sprawling campus in Bethesda, Maryland. Graham's two superiors, John Mascola, director of the Vaccine Research Center, and Anthony Fauci, director of the National Institute of Allergy and Infectious Diseases, were there as well. They huddled around a narrow conference table near Fauci's office, with Bancel sitting directly across from Fauci.

This was the moment of truth. For many, Bancel had a reputation of being more of a salesman than a scientist. The Nipah project wasn't going to be a moneymaker. The others in the room wanted to know if Bancel was really interested in a project that might never generate any profit for his company. Was he really committed to vaccines for an infectious disease? And was he really interested in testing out the concept of rapid response to an outbreak? Bancel, a Frenchman who despite years working in Cambridge, Massachusetts, still has a thick accent, assured Fauci and the others that he was committed. However, to satisfy his shareholders, he would need at least enough money to cover his expenses. Mascola later called that meeting "pivotal."

Now it was just a question of when to click the stopwatch. But before that happened, fate stepped in. On New Year's Eve, fragmented news reports came out of China about the outbreak of a respiratory illness in Wuhan, a city the size of London 500 miles west of Shanghai. The city's health commission said 27 people were sick with viral pneumonia. Seven were in serious condition.[6]

Seventeen years earlier, China was the site of an outbreak of the novel coronavirus SARS, an acronym for severe acute respiratory syndrome. That virus had leaped from bats to civets, nocturnal mammals resembling raccoons, before making the leap to humans. The SARS outbreak killed 800 people, about 15% of those with confirmed cases. Half of those age 65 or older died.[7] Fortunately, SARS, though lethal, was not highly contagious. But the Chinese government endured blistering criticism for not revealing more about the origins of that virus.[8]

Less than a week after those New Year's Eve reports, the patient count was up to 59. Authorities had shut down the Huanan Seafood Market and sent in workers wearing hazmat suits to disinfect it, including the stalls where wild meat had been sold. Public health officials were still baffled, at least publicly, but they were starting to rule out known diseases, such as SARS, MERS, and the bird flu.[9] Graham strongly suspected it was a new strain of coronavirus. He began preparing, just in case.

He called a former colleague, Jason McLellan. At the time, the 38-year-old scientist from the University of Texas was getting new snowboard boots refitted at a ski shop in Park City, Utah. The new boots were horribly uncomfortable, so the shop was remolding them with hot air guns. McLellan had worked closely with Graham on a breakthrough on the RSV vaccine and later a MERS vaccine. Graham wanted to know if he was game for helping on a vaccine for whatever virus was spreading in China. McLellan was enthusiastic.

That same day, January 6, Graham met with Kizzmekia Corbett, a research fellow in his lab. Corbett had worked closely with Moderna to formulate a MERS vaccine in mice. They mapped out a strategy for what to do when the genetic code for the virus was released. From their previous work, they already had a good sense of what would go into the vaccine. Corbett's job then would be to test that new vaccine in mice to see if it produced the right antibodies. Only then could it be tested in humans. She was not going to have much time. The goal was to start a human trial within 100 days.

Meanwhile, Bancel was vacationing at his home in the south of France when he read about the virus in the *Wall Street Journal*. He sent a link of the story to Graham by email, asking him if he could make sense of it.[10] Graham shot back an email: "If it's a SARS-like coronavirus, we know what to do and have proven that mRNA is effective at a very low dose. We were waiting to have verified sequences before I called, but this would be a great time to run the drill for how quickly can you have a scalable vaccine."

Graham had made a fateful decision. He would scrap plans for a Nipah vaccine and focus instead on the new virus spreading in China. He had no idea whether the virus would be highly contagious. MERS and SARS were not. He also had no idea how far the virus would spread. But the game had changed. The Nipah vaccine project had been mostly a demonstration. He wanted to prove to the scientific community that he could get an experimental vaccine into a human trial in world-record time. He didn't give up that pursuit. But now the project was very different. He wanted to make an actual vaccine that could be approved by the FDA and given to the public. That would take a lot more money than he had.

Graham met with Mascola and Fauci at the same conference table where they had felt out Bancel. This was before China had released the virus's genetic sequence. Graham told Fauci, "Just get me to sequence. Just get me to the sequence and we're on the road." Fauci didn't have to sign off on switching to a coronavirus vaccine. "Even though I'm quote-unquote his boss, it's really very much of an egalitarian type of collaboration," Fauci would explain later in an interview for this book. Graham would outline to Fauci in detail the significance of the research he had done in previous years. Fauci was aware of the research, but he did not realize just how much it had altered the course of vaccine science or how critical it suddenly was. Graham had such a head start on a coronavirus vaccine that he convinced Fauci he could deliver it within 18 months, a prediction Fauci would soon spread in the national media. Graham said, "Let's go full blown. Let's make a vaccine." He asked Fauci if there was enough money. Fauci didn't hesitate: "Barney, let me worry about the money. Go ahead and do it."

Graham woke up the following Saturday morning with the sequence in hand, ready to make a vaccine. It was unseasonably warm for January in the Washington, DC, area. The *New York Times* that day reported the first death from the virus: a regular customer of the seafood market.[11] Graham grabbed a cup of coffee and walked over to his office at the front of his Rockville, Maryland, home. The office had jam-packed floor-to-ceiling bookcases. Among the items on the shelves were wooden carvings of the words *Faith*, *Hope*, and *Love*.

The most critical decision to be made was the easiest. Everyone knew from years of prior work exactly what would go into the vaccine. It would build off years of research of a MERS vaccine that McLellan and Corbett had helped develop by 2017. They would use mRNA to trick human cells into producing a very precise version of the spike protein on the surface of the virus. They would use a special technique to stabilize that protein, a technique they had already filed a patent for in 2017. They would use two protein fragments called prolines to lock the spike protein into the same shape found on the virus before it attacks a human cell. That was the key. Corbett still marvels at how they knew the answers instantly. Had they not spent years already studying

MERS, there is no way they could have turned around a vaccine so quickly. There were other more technical decisions to make. Supplies needed to be ordered. Moderna would have to be told so they could make the vaccine. Corbett needed to start preparing for the mouse studies.

Graham spent the morning exchanging texts, emails, and phone calls with Corbett, McLellan, and a few others. He also talked to his boss, Mascola, and his boss's boss, Fauci, who signed off on the plan.

With Moderna's technology, Graham didn't need the virus itself, a radical departure from the standard way of making a vaccine. Most vaccines come from a live virus that has been killed or rendered harmless, or from a part of the virus. Instead, Graham could analyze the code on his laptop, using a variety of software, and design a vaccine digitally. Worried about a rush of demand for the precise proteins he needed to study, he ordered them from a laboratory within hours.

In his mind, he had clicked a stopwatch to see how fast he could get the vaccine he created into human trials. That meant meeting with Moderna on Monday to get them to produce the vaccine. Then he had to set up trials in the laboratory, in mice, and later in monkeys, all before giving the vaccine candidate to humans.

Although Moderna would eventually get most of the credit, it was NIH that led the way in the early weeks. Mascola now admits that he didn't realize the virus was so contagious until February, with news of the virus spreading on the cruise ship *Diamond Princess* that had departed Hong Kong. Ultimately, 700 of the 3,711 people aboard became infected—a sign that this disease could be a serious health threat. Once hospitals in Italy were overwhelmed, Mascola knew we had a pandemic on our hands.

Just about the time schools across America started shutting down in mid-March, Moderna's demonstration vaccine was given to the first volunteer in Seattle. It had been done in a record-setting 66 days. And by then, it wasn't just a demonstration. The virus was spreading. The country went into a lockdown. The economy was collapsing. Everything had changed. The world was now eagerly waiting for the vaccine so that it could return to normal.

CHAPTER 2

History of Vaccines

In October 1977, a man transporting two smallpox-infected children to an isolation camp in Somalia got hopelessly lost. The driver stopped at a hospital near Mogadishu to ask for directions. A 23-year-old cook who worked there, Ali Maow Maalin, jumped in the van and offered to show the way. Before they took off, the driver turned to the new passenger to ask if he had been vaccinated against the highly contagious smallpox.

"Don't worry about that," Maalin said. "Let's go."[1]

Maalin in fact wasn't vaccinated. Even though he had previously worked for a smallpox eradication program, Maalin was so afraid of needles that he'd hold his arm and pretend that he'd already received his skin pricks. Within minutes of the van ride, Maalin was infected.

Luckily the virus he caught, variola minor, causes less severe illnesses than the more common variola major form of smallpox.[2] Even more fortunately, Maalin's hospital moved quickly to isolate him and to quarantine and vaccinate exposed patients and staff. With nowhere to spread, the virus quickly petered out.

Maalin eventually recovered. He didn't know it then, but he was the last person in the world to catch naturally occurring smallpox. The last vestiges of the smallpox virus couldn't find another host in which to replicate and survive. Two years later, in 1980, the World Health Organization declared the virus eradicated. It remains the only human disease we have managed to wipe out. The World Health Organization hopes to do the same with the polio virus within the next few years.

Vaccines have revolutionized modern medicine. At the turn of the 20th century, the life expectancy of the average American was 47.3 years. For Black Americans, it was only 33 years.[3] But a combination of improved health care and public health measures, such as clean water, combined with vaccines for smallpox, whooping cough, diphtheria, and tetanus, helped to dramatically increase longevity. One estimate is that since 1924, vaccines have prevented 40 million cases of diphtheria and 35 million cases of measles, both deadly viruses. That's on top of 103 million cases of other childhood diseases prevented.[4] As a consequence, the life expectancy of the average American today is 78.7 years. For Black Americans, it is now 75.3 years, according to the Centers for Disease Control and Prevention.[5]

The early vaccines were a product of some luck. Smallpox's disappearance came nearly two centuries after an English physician named Edward Jenner scientifically established the concept of vaccination. Specifically, Jenner was the first to document that if you inject people with pus from smallpox's milder cousin, cowpox, they can be protected against the more lethal virus.

Smallpox is a terrible, deadly ancient disease. It spreads easily through respiratory droplets and contaminated surfaces. It killed about one in three infected people.[6]

Historians believe that 90% of the native population of the Americas succumbed to smallpox after exposure from European explorers. Nearly one-third of those who survived the virus went blind, and many were left with disfiguring scars.[7]

Doctors in China as far back as the 10th century tried to inoculate people against smallpox by swabbing their noses with[8] fluid from smallpox pustules or pieces of dried scab. This process, called variolation, was built on the knowledge that people who survive smallpox don't get it again. It may have been a lucky guess. Viruses wouldn't be discovered for nearly another 1,000 years.[9] Clearly, no one could have understood the biological mechanisms behind variolation.

Variolation could be a dangerous game of chance: some subjects got a fever, rash, and other smallpox symptoms. Others died, although in smaller numbers than if they had caught smallpox naturally.[10] And variolation sometimes backfired horribly and set off smallpox epidemics.[11]

Despite the risks, George Washington ordered all his troops to be inoculated on February 5, 1777. Washington, who survived smallpox as a teenager, considered it potentially more deadly than British forces.[12] "Finding the smallpox to be spreading much and fearing that no precaution can prevent it from running through the whole of our army, I have determined that troops shall be inoculated," he wrote in a letter to John Hancock, the president of the Second Continental Congress.[13]

Years later, Jenner's alternative method laid a safer foundation for immunology. He had witnessed that milkmaids who became sick with cowpox didn't seem to contract smallpox. Jenner's first unwitting subject was a healthy eight-year-old child named James Phipps. According to historical records, Phipps was either a pauper or the son of Jenner's gardener. Whoever he was, the boy obviously was too young to consent to the experiment.[14]

On May 14, 1796,[15] Jenner infected Phipps with scrapings from cowpox lesions taken from the hand of a milkmaid named Sarah Nelmes. The boy got mildly sick, had a fever, and lost his appetite. But 10 days after his inoculation, Phipps felt much better. Two months later, Jenner infected Phipps again, this time with a fresh smallpox lesion. Young James didn't get smallpox, a sign that his earlier exposure to cowpox had rendered him immune.[16]

Jenner submitted a brief paper on his findings to the Royal Society. For reasons unknown, the editors rejected it. So the following year, in 1798, Jenner privately published a booklet titled "An Inquiry into the Causes and Effects of the Variolae Vaccinae, a disease discovered in some of the western counties of England, particularly Gloucestershire and Known by the Name of Cow Pox."[17] *Variolae* is Latin for pustules. *Vaccinae* is derived from Latin for cow.[18] In what may be his most lasting scientific contribution, Jenner named the new procedure vaccination.[19] Parliament banned variolation in 1843 and later made the cowpox vaccine mandatory.[20]

But it wasn't until the late 1800s that the world had its first true vaccine.[21] And it happened thanks to a negligent laboratory assistant.

In 1879, French chemist Louis Pasteur left on a holiday after instructing his assistant to inject a batch of chickens with fresh cultures of *Pasteurella multocida*, the bacterium that causes chicken cholera. The assistant forgot before he departed for his own holiday. The untended brew sat in the lab for a month, plugged with just cotton wool.[22]

When Pasteur realized the mistake, he might have just tossed the bacteria. But instead, he had his assistant go ahead and belatedly give the chickens the unrefreshed cultures. Pasteur was shocked by the results. Instead of dropping dead of cholera, the animals showed mild symptoms and then recovered. On a hunch, Pasteur decided to inject the birds with fresh cultures of *P. multocida*. Again, rather than die, the chickens stayed healthy. Pasteur theorized that exposure to oxygen had weakened the cholera bacteria—it was potent enough to sicken the birds but not enough to kill them. He also reasoned that the chickens' immune defense had been primed by the enfeebled bacteria to fight off a full infection. Pasteur called the progressive loss of virulence after exposure to air "attenuation."[23]

Thus the world had its first lab-made vaccine.

Pasteur had previously proven that diseases were caused by visible "germs," a notion that his critics considered preposterous.[24] But the discovery is one reason why some regard Pasteur as the pioneer in microbiology and immunology. Now he aimed his scientific focus on attenuation to vaccinate against a host of other diseases. Along with Charles Chamberland and Émile Roux, Pasteur developed a live vaccine for anthrax for livestock. He showed that anthrax spores cultivated at high temperatures became less virulent.

In 1881 in Pouilly-le-Fort, France, Pasteur orchestrated a public experiment with gathered journalists by inoculating 24 sheep, six cows, and one goat with a weakened *Bacillus anthracis* culture. On May 31, those animals plus a control group of 24 sheep, four cows, and a goat were exposed to freshly isolated anthrax pathogens.

The vaccinated animals were unharmed. All the unvaccinated sheep and the unvaccinated goat had dropped dead; the four unvaccinated cows showed symptoms of anthrax. Pasteur proclaimed the demonstration an "éclatant succès," a resounding success.[25]

That same year Victor Galtier, a French veterinarian, discovered that sheep injected with saliva from rabid dogs not only don't get rabies, but also are protected from future infections.[26] Galtier's surprise findings spurred Pasteur and his collaborators to go to work to develop a vaccine for rabies, which was 100% fatal in humans.

Pasteur hit a roadblock right away. He couldn't culture the pathogen in the lab or even examine it under a microscope. That's because unlike anthrax, rabies is caused not by bacteria but by viruses.

But Pasteur and his partners soon learned that passing the virus between different species of animals turned them progressively less virulent. So inoculating a sample from a rabid dog into a monkey attenuated the virus. That's the opposite of what happens if rabies is passed from the same animal type to the next, when it serially grows more virulent.

What's more, every subsequent interspecies transmission further weakened the virus. So with repeated transmissions among dogs, monkeys, rabbits, or guinea pigs, Pasteur could tailor the degree of virulence needed for a safe vaccine for dogs.[27]

Pasteur air-dried spinal cords from infected rabbits after passing the rabies virus taken from dogs through multiple rabbits. In an 1885 report to the French Academy of Science, Pasteur credited exposure to oxygen with attenuating the virus. He successfully vaccinated 50 dogs against rabies. Today we know that the attenuation actually resulted from passing the virus through dissimilar species.[28]

On July 6, 1885, a reluctant Pasteur, who was not a physician, tested his rabies vaccine on his first human subject. The patient was nine-year-old Joseph Meister, who had been attacked by a neighbor's rabid dog in Alsace two days earlier.

Over 11 days, the boy was inoculated 13 times with the emulsion from the spinal cord of a rabbit that had died of rabies. Each time the vaccine came from progressively fresher—and therefore more virulent—part of the cord. After three months, doctors declared young Joseph fully recovered.

In 1886 alone, Pasteur went on to inoculate 350 people against rabies in Europe, Russia, and elsewhere. Just one person developed rabies. The method was initially called "Pasteur's treatment."[29] But in an homage to Edward Jenner, Pasteur gave the artificially weakened microbes given to trigger the body's natural immune defenses the generic term *vaccines.*

Pasteur's advances sparked the nascent field of virology. Five more vaccines followed during the first half of the 20th century: diphtheria, pertussis, tetanus, tuberculosis, and influenza.[30]

The vaccines that rolled out during the next 50 years aimed to banish some of public health's biggest menaces: polio, measles, pneumonia, hepatitis A and B, meningitis, and chicken pox.[31]

Creating effective vaccines requires a mixture of science, hunches, and luck. Failures are a given. Well into the 20th century, searches for vaccines sometimes sickened, maimed, or even killed human subjects.

Scientists tested unproven vaccines on both eager volunteers and unwitting subjects. They inoculated children who were wards of the state, including those with mental retardation; enlisted service members; their own offspring; and even themselves. Researchers justified the risks as a price paid to fight the greater dangers posed by infectious diseases.

On March 6, 1857, a *New York Times* article on commencement exercises at New-York Medical College reported that Nehemiah Nickerson won first place "for his thesis on Infantile Paralysis." That was the first appearance in the newspaper of what later became known as poliomyelitis, or polio.[32]

The *Times* made little mention of the disease for the next four decades. Then on August 7, 1899, polio suddenly was front-page news. The disease was "spreading with remarkable rapidity," with cases popping up in Poughkeepsie, New York, and nearby towns. But once again, infantile paralysis largely disappeared from news pages for the next decade.

Polio probably has plagued humans for millennia. An Egyptian carving dated to 1400 BCE depicts a young male with a leg deformity likely caused by polio. It circulated at low levels through the 1800s. By 1910, however, infantile paralysis was raging in the US and other well-off nations.[33]

In 1905, Swedish physician Ivar Wickman recognized that poliomyelitis was a contagious disease that passes from person to person.[34] Polio is caused by an enterovirus that enters through your mouth and lives in your throat and intestines. You get it mainly through contact with an infected person's feces.[35] Most people have no or mild symptoms. But polio can lead to paralysis and even death.

Polio's resurgence in the 1910s kick-started a decades-long hunt for an effective vaccine. In 1932, John Kolmer at Temple University in Philadelphia developed one of the first candidates for the polio vaccine after watching his colleagues' successful work on denatured vaccines for rabies and yellow fever. Kolmer theorized that he could weaken the polio virus just enough to trigger protective immunity without risking a full-blown disease. He ground up the spinal cords of polio-infected monkeys, soaked them in a salt solution, filtered them through a thin mesh, and then treated them for 15 days with ricinolate, a substance found in castor oil. He tested the vaccine on his two sons, his lab assistant, and 25 local children.

Kolmer then tested his three-dose vaccine on thousands of American and Canadian kids. Several children contracted polio shortly after receiving their shots. Paralysis started in the inoculated arm and spread. In all, Kolmer's vaccine paralyzed 10 children and left five dead.

Over at New York University, a young researcher named Maurice Brodie took a different path to a vaccine. Instead of using a weakened but live virus as Kolmer had, Brodie killed the poliovirus taken from the spinal cord of infected monkeys by soaking it in a 10% solution of formaldehyde for 25 days.[36]

Brodie tested his vaccine on chimpanzees, himself, five colleagues, and a dozen children. When they produced polio antibodies without getting the disease, Brodie went on to enroll some 11,000 children in a trial to prove the efficacy of his formaldehyde-inactivated vaccine. But again, some of the children became paralyzed or died. Critics denounced Brodie's study as poorly run and his vaccine as useless.[37]

Poliovirus researchers faced other challenges, too. They relied on infected monkeys for research, specifically to harvest their spinal cord and brain for the poliovirus. But that was expensive and laborious and required sacrificing the lives of tens of thousands of primates.[38] Researchers spent fruitless decades to find a way to grow the poliovirus outside the bodies of live monkeys or other animals.

Then a major breakthrough followed in 1949. Thomas Weller, John Enders, and Frederick Robbins were working on the chicken pox virus in Boston when they chanced on a discovery: they could grow poliomyelitis viruses in test tubes in the lab, and with skin and muscle tissues from human embryos. Researchers no longer needed to rely on live animals as incubators—just their nervous tissue.[39]

The tissue culture technique also meant that vaccine stocks could be produced in big enough volumes to inoculate millions of people. The three American scientists were jointly awarded the 1954 Nobel Prize in Physiology or Medicine for their pivotal first step toward a vaccine for polio.[40]

Two years before the discovery by Enders, Robbins, and Weller, a young virologist named Jonas Salk joined the University of Pittsburgh to set up his virus research lab.[41] Salk had come from the University of Michigan, where he had been studying the influenza virus under his mentor, Thomas Francis Jr.

Salk was the eldest of three sons born to Russian-Jewish immigrant parents in New York City. He was academically precocious, driven, and egotistical. Salk's parents, who worked in the garment industry, planned for him to attend rabbinical school. But over his mother's objections, Salk wanted to study law and run for Congress. However, he discovered chemistry while at the City University of New York. Then as a medical student at New York University, Salk redirected his focus to biological research.[42]

Thirteen years before Salk graduated from CUNY in 1934, 39-year-old Franklin D. Roosevelt contracted polio, which paralyzed both of his legs below the knee. A year before Salk earned his medical degree in 1939, Roosevelt founded the National Foundation for Infantile Paralysis, now known as the March of Dimes, to combat polio. Roosevelt's close friend and former law partner

Basil O'Connor became the nonprofit organization's driving force for the next three decades.

Salk believed that a vaccine made of killed virus particles—which Maurice Brodie had tested earlier to disastrous results—could work as well as vaccines made from live but weakened viruses. Some researchers were skeptical that lab-altered vaccines could elicit the body's immune response.[43]

In September 1951, Salk and O'Connor met aboard RMS *Queen Mary* while sailing home from the Second International Poliomyelitis Conference. O'Connor was intrigued by Salk's theory that exposing the polio virus to formaldehyde could produce a vaccine that was both effective and safe. O'Connor decided to back Salk's research with the foundation's money.

"Before that ship landed I knew this was one young man to keep an eye on," O'Connor recalled.[44]

Salk had learned how to inactivate viruses with formalin, a solution made with formaldehyde, during his studies on flu. He turned to the same technique for polio. Salk grew the polio virus in monkey kidney cells, then killed them with formalin. Two years after his meeting with O'Connor aboard the ocean liner, Salk had a vaccine he was sure would work. In the spring of 1953, Salk injected his whole family: his wife, Donna; his three young boys; and himself. The decision, he said, was "courage based on confidence, not daring." He wanted his family to be among the first to be protected against polio.[45]

The youngest son, Jonathan, was three when he was photographed by the March of Dimes as his father plunged the needle in his left arm as his mother protectively held both of his hands. The boy, wearing a jaunty bowtie, bravely seems to be stifling a cry.

Jonathan grew up to become a physician. As an adult, he recalled that his father "wasn't experimenting on us. He gave us the vaccine because he knew it worked, and he wanted us protected. There was, however, an element of fear and a kind of trauma in getting the shot from him."

At a press conference at Midtown Manhattan's Waldorf-Astoria Hotel that November, Salk revealed the successful experiment on his family. "I will be personally responsible for the vaccine," Salk said with a typical brashness that rankled more than a few colleagues and competitors.[46]

Soon after, a mammoth placebo-controlled field trial began to validate Salk's initial results. It was run by Salk's former professor, Thomas Francis Jr., at the University of Michigan School of Public Health. The study ultimately enrolled 1.8 million American, Canadian, and Finnish children and lasted two years.[47]

When Salk arrived at the University of Michigan for the big reveal of the study's results, he was still in the dark himself. But they told him the morning of April 12, 1955, that the vaccine was up to 90% effective. It was a huge relief for Salk.[48] The rest of the world would find out shortly, when Francis announced to 500 scientists and physicians gathered inside University of Michigan's Rackham Auditorium that "the vaccine works. It is safe, effective, and potent."[49] After the press conference, CBS reporter Edward R. Murrow asked Salk who owned the vaccine. "Well the people, I would say," Salk said. "There is no patent. Can you patent the sun?"

The vaccine turned Salk into something of a national messiah. Dwight H. Murray, chairman of the board of directors of the American Medical Association, called the news "one of the greatest events in the history of medicine." President Dwight D. Eisenhower later hailed Salk as a "benefactor of mankind."[50]

Salk's vaccine promised to turn the tide on a disease that consigned some victims to life in wheelchairs and tanklike breathing machines called iron lungs. In one year, 80 million people donated money for Salk's research, in dimes and dollars.[51]

As soon as Salk licensed his vaccine, American officials began mass inoculations. But a disaster struck almost immediately. Batches of vaccines made by Cutter Laboratories in Berkeley, California, were filtered improperly and as a result were contaminated with live polio viruses. Salk's vaccine used three virus strains: MEF-I, Saukett, and the highly virulent Mahoney strain.[52] Some of the Mahoney virus strain had evaded inactivation by formaldehyde, possibly because some virus particles had clumped together. Some 200,000 children received the tainted vaccines. As many as one-third of them contracted polio. Of those, 200 were paralyzed to varying degrees and 10 died.[53]

The US Surgeon General ordered all polio vaccinations stopped to investigate manufacturing protocols. The probe led to the addition of a second filtration step and improved safety tests. Vaccinations resumed that fall.

The Salk vaccine had several early drawbacks: It required three separate shots from a nurse or a health professional, injected just below the skin. Inoculated children lost some of the antibodies against polio after a few years. And producing the vaccine was cumbersome and inhumane; for every 1 million inactivated doses, 1,500 monkeys were killed for their kidney cells.[54]

But the Salk vaccine was indisputably effective. In 1962, the number of polio cases in the US plummeted by 98 percent compared to before the vaccine was widely available, to just 910 people.[55]

Even as Salk was being hailed as a public health hero, other researchers were pursuing what they considered a superior option: live attenuated vaccine. Among them was Hilary Koprowski, a

Polish-Jewish physician who arrived in the US in 1944 to flee the Nazis.

Koprowski spent most of his career in private industry, not in academia, which was unusual among prominent virologists at the time. It was at Lederle Laboratories in New York that Koprowski, in January 1948, poured a thick, gray, and greasy brew from a kitchen blender into a beaker and drank it. The cold cocktail—which Koprowski said tasted like cod liver oil—contained mainly cotton rat brain and polio virus that he had weakened in his lab. With the sip, Koprowski became the first person to be inoculated with a live-virus polio vaccine. Two years later, Koprowski carried out the first successful trial of a weakened virus vaccine.[56]

Today, Koprowski's landmark feat is largely a historical footnote. He was eclipsed by Albert Sabin, a fellow Polish immigrant physician who also believed that the best way to defeat polio was with a live vaccine. Sabin began working on polio virus in 1931 as a young researcher at New York University.

Sabin thought that any vaccine would need to pass through the same route as wild polio viruses: through the mouth into the intestine.[57] Sabin discovered that chimpanzees were the best animal species to test gut infectivity. He also found that rhesus monkeys and cynomologous monkeys, also called long-tailed macaques, made the best primate subjects for testing nervous systems.

The field of virology was largely split into two battlegrounds on polio vaccine—using inactivated virus or the more conventional live attenuated virus. It also was riven by professional jealousies, personal attacks, and public sniping. When Salk published his pathbreaking paper in the March 28, 1953, issue of the *Journal of the American Medical Association* to report that he had successfully vaccinated 161 people, Sabin trashed the findings. He warned that Salk's use of monkey kidney tissue could cause people's own immune systems to damage their own organs.[58] Sabin years later dismissed Salk's vaccine as "pure kitchen chemistry. Salk didn't discover anything."[59]

Near the end of his life, Salk accused Sabin of being driven by professional enmity. "Albert Sabin was out for me from the very beginning," he said in a 1993 interview. "In 1960, he said to me, just like that, that he was out to kill the killed vaccine."[60]

Because Salk's vaccine was licensed first and used extensively in the US, Sabin in the late 1950s was forced to conduct his clinical trials overseas, in the Belgian Congo and the Soviet Union. A key advantage of Sabin's vaccine, which was given three times by mouth with sugar cubes, was that people shed weakened virus immediately through their excrement. Since polio spreads mainly through contact with feces, the live vaccine boosted herd immunity.

In 1960, after studies of several live weakened polio vaccines from Sabin, Koprowski, and others, the US surgeon general, Leroy E. Burney, selected Sabin's as slightly safer than Koprowski's. Sabin's vaccine received its first license next year. Many researchers felt that Koprowski's earlier work with a live-virus vaccine paved the way for Sabin's achievement. Koprowski himself told his biographer, Roger Vaughan, that "sometimes I introduce myself as the developer of the Sabin poliomyelitis vaccine."[61]

Sabin's oral vaccine was cheaper and easier to administer than Salk's needle-based vaccine. In 1962, Cuba launched the first mass vaccination campaign with the Sabin vaccine. The United States switched from Salk's vaccine in favor of Sabin's oral vaccine. The rest of the world, too, ditched the Salk vaccine for the next four decades.[62]

But Salk never gave up championing his killed vaccine. He repeatedly pointed out that, though extremely rare, live vaccines can grow strong enough to cause polio and paralysis. Unvaccinated parents, for instance, were infected while changing dirty diapers. In the United States and other countries that have eradicated polio, the risks from live-virus vaccines actually outweigh any benefits. That's why the US in 2000 discontinued the use of Sabin's vaccine and returned to Salk's old-fashioned shots.[63]

Polio was just one of several key vaccines that rolled out during the 1950s thanks to viral tissue culture techniques. Others include vaccines against measles, mumps, rubella, and chicken pox.[64] In all, we now have effective vaccines against at least 28 human pathogens. Among the newest inoculations are for shingles, the sexually transmitted human papillomavirus (HPV), and the deadly Ebola virus, many of whose victims bleed to death from burst cells and leaking blood vessels.[65]

But despite decades of research, some major vaccines have defied invention. The most elusive of all is the one against HIV. Until recently, we lacked a vaccine against malaria, a parasitic tropical disease to which half of the world's population is vulnerable and which kills some 400,000 people each year.[66]

One reason for the failures is that making good vaccines is difficult work that can take decades. For instance, we've known since 1880 about the malaria pathogens that infect mosquitoes that feed on humans.[67] One of the fastest vaccines ever developed was for measles. It took only 10 years.[68] But that record was broken in the 1960s with a mumps vaccine that took only four years to get approved[69]—a record that was shattered in 2020 by the blazing speed of vaccines for COVID-19.

In 1966, one vaccine trial for respiratory syncytial virus (RSV) ended catastrophically after[70] many of the children in the study ended up in the hospital and two toddlers died.[71] The technique used for that vaccine was very similar to the one used by Jonas

Salk. It taught vaccine hunters that different diseases responded differently to the same approach.

In the 1970s, a new, safer approach to vaccines was introduced. Instead of killing or weakening a whole virus, genetic engineering allowed researchers to use just a small part of the virus. That made it impossible for human subjects to get infected and suffer from the actual disease. One of the first vaccines created with the approach was for whooping cough. By using just a component of the bacteria, the vaccine had fewer side effects. A vaccine for HPV relied on a protein found on the virus's outer shell.[72] But even protein-based vaccines didn't eliminate the possibility of vaccines making a disease worse. That was a dilemma Barney Graham would toil away at for most of his career.

CHAPTER 3

The Underrated Scientist

In the world of science, DNA is a superstar. From the discovery of the double helix to cracking the code to CRISPR, deoxyribonucleic acid has been at the center of some of the greatest biological discoveries. Scientists cannot go wrong devoting their careers to understanding the blueprint of life. While we all know that DNA determines the color of our eyes, our hair, and our skin, medical researchers still believe that—despite setbacks—manipulating genes will one day enable us to cure many diseases.

But until recently, RNA seemed like an extra on a movie set. Messenger RNA was discovered in 1961, but scientists could not do anything with it until 1984 when Paul Krieg and Douglas Melton published a recipe for how to make mRNA in an obscure journal.[1] Melton went on in 1987 to help start Gilead Sciences, a California biopharmaceutical company that would become one of the most successful makers of antiviral drugs. Melton's idea was to use mRNA to block the replication of viruses. However, Gilead never developed that technology. While mRNA attracted some interest from early biotech companies in the 1990s, attempts to make it do anything useful appeared fruitless. By

2000, very few scientific labs were taking RNA seriously as a possible therapeutic.

This was true even though DNA's true power comes from RNA—messenger RNA, to be exact. Our lives depend on DNA making RNA, which then makes proteins. This happens constantly in the 37 trillion cells in our bodies.[2] Tightly coiled DNA is trapped inside the nucleus of our cells. The double-stranded helix works by creating messenger RNA, a single-stranded replica of a gene. Messenger RNA can slip through the pores of the nucleus into a cell's inner world, known as cytoplasm. Ribosomes find the RNA and, much like a 3D printer, read the code on mRNA to link amino acids together to form proteins. We tend to think of proteins as something we consume in meat or cheese to help build muscle. Vegans are constantly asked if they are getting enough protein. But our bodies produce their own proteins—more than 20,000 different varieties—to do just about every task imaginable. Proteins help new cells grow. Proteins like insulin regulate body functions. Antibodies are made up of proteins. So are enzymes. In short, mastering mRNA offers the possibility of giving our bodies the genetic code to heal themselves.

If a person has a defective gene, they might not be able to produce a protein that the body needs. Sickle-cell anemia, for example, is a common disease caused by a defective gene handed down at birth. That defect leads to sickle-shaped red blood cells that can be more easily destroyed. Because red blood cells carry oxygen throughout the body, a lack of functioning cells leads to anemia.[3] RNA is the messenger from the DNA that produces the protein molecule hemoglobin. Giving a sickle-cell anemia sufferer the correctly coded RNA would produce normal hemoglobin, fixing the problem at least temporarily. Just how temporary the solution would be remains a big unknown. But in theory, at least, RNA could offer an alternative to gene therapy.

Who came up with the idea of using mRNA as a drug first? No one knows. "You're not going to find a person who had that idea first," said Melton, himself a pioneer in mRNA research. He said early discoveries made it obvious, even though scientists lost interest in mRNA over time.

One of the few scientists obsessed with the potential for mRNA was Katalin Karikó. She devoted her career to trying to understand RNA, an infatuation that ironically made her a highly underrated scientist. If a professor teaches history or English, universities pay them a decent salary. But if the professor is a scientist, they are expected to bring in money themselves through grants. The old saying "publish or perish" applies especially to scientists working at research universities. The top-tier scientists publish in the most elite medical journals and get six-to-seven-figure grants. That can be difficult if the researcher pursues a topic in which the rest of the scientific community has lost interest. For many, RNA was seen as a lost cause. Kati, as her friends call her, failed time and again to find funding to continue her research and struggled to hang on to university jobs. "Every night I was working: grant, grant, grant," Karikó remembered. "And it came back always no, no, no."[4] It was a predicament that would define most of her career.

Karikó is a fast-talking extrovert with amazing recall. A conversation with her has the feel of an amusement park ride, with a rush of details and scientific jargon coming through her thick Hungarian accent. Standing six feet tall, she has an imposing presence, and she is not afraid to speak candidly to her peers. She credits her resilience to a book her biology teacher made her read in high school. It was written by famed Austrian endocrinologist

Hans Selye, who studied the effects of stress. The lesson that Karikó remembers from the book is "You have to focus on what you can do and don't focus that much on what others are saying."

Karikó was born in Hungary, just shy of two years before Soviet leader Nikita Khrushchev brutally crushed a 12-day uprising in Budapest with tanks and troops. Thousands were killed or maimed.[5] Karikó's parents, who had endured Nazi occupation, went on to live under Soviet domination. Yet, Karikó grew up in Kisújszállás, a town small enough and out of the way enough to escape the brutality.

"I was a happy girl," she recalls. "My father was a butcher, and I liked to watch him work, observe the viscera, the hearts of the animals. Perhaps that's where my scientific vein came from."[6]

Karikó's introduction to working with RNA was a sheer coincidence. One day as a student at the University of Szeged, Karikó was in an office when a scientist strolled in, looking for students interested in working on his RNA project. Karikó had spent time on another project at the Biological Research Center feeding fish corn to analyze what types of fat molecules they would produce. (It taught her never to eat farmed fish again.) Those molecules, called lipids, can be used to create liposomes, essentially fat bubbles that can deliver encapsulated DNA to the nucleus of the cells of mammals. Colleagues recommended Karikó to the scientist.

Karikó married a young engineer, Béla Francia, in 1980. Two years later, she earned her doctorate degree and gave birth to their daughter, Susan Francia, who would later become a two-time Olympic rowing gold medalist at Beijing and London. She and her husband loved their lives in Hungary, but the school couldn't find funding for her research. For a country behind the Iron Country, Hungary had flourished, escaping the poverty plaguing Poland and Yugoslavia. But by 1985, the economy there was stagnating.[7] Karikó learned on her birthday in January 1985 that she would lose her job by July. She was desperate to stay in Europe, but the only job she could find was in Philadelphia, at Temple University.[8]

So her family packed up their belongings, sold their car, and secretly stashed about $1,200 in her daughter's teddy bear to get past Hungary's strict laws on taking money out of the country.[9] Karikó ripped the seams off the teddy bear and stitched them back up herself.[10]

She had a two-year contract for $70,000 a year at Temple. Not bad, but it was a step back for her. Her husband, Béla Francia, had to give up his job as an engineer and settle for menial odd jobs. He painted houses, cut grass, and even worked as a janitor. He eventually managed an apartment complex. Karikó's salary as a postdoctoral researcher had to support the family of three, as well as her mother-in-law. Simple tasks such as washing clothes now meant trips to a coin-operated laundromat in the basement of their apartment complex.

She worked on a clinical trial at Temple, using double-stranded RNA on HIV patients, but it did not go well. Still, she recalled, "I worked day and night and I was very driven." In 1988, she got an offer from Johns Hopkins University, which made her boss at Temple irate. Karikó said he even threatened to have her deported, leveraging the power of Temple having sponsored her work visa. Johns Hopkins withdrew the offer over complications in her immigration status. Karikó was now determined to leave Temple no matter what. She couldn't forgive her boss for the way he had threatened her.

She landed a job the next year at Uniformed Services University of the Health Sciences, a military school to train doctors, nurses, and dentists located in Bethesda, Maryland. Her family stayed in their apartment in Philadelphia. That meant that Karikó would head out at 3 a.m. each Monday to drive 140 miles to Bethesda and return late on Friday nights. She never had an apartment in her nine months there. She often slept in her office. Or she would crash at a friend's place, slipping in after they had gone to bed and leaving before they woke up. Sometimes the only way they knew she had been there was because the newspaper was not in the yard but was on the kitchen counter instead.

It was a difficult time. Karikó was spending lots of money on a lawyer to fix her visa issues so she could stay in the US. The long commutes drained her, as did all the weeks she spent alone. She finally managed in 1989 to find a low-level research assistant job at the University of Pennsylvania working under a young cardiologist, Elliot Barnathan, who years later would become the executive director of research and development at the pharmaceutical company Janssen. Karikó's job didn't pay well and wasn't on a tenure track. Given how unattractive the terms were, only foreigners would take those jobs, Karikó recalls, adding, "I was desperate."

The real excitement in medical research emanated from the $3 billion Human Genome Project, the seemingly fanciful attempt to map all genes in our DNA. Led first by Nobel laureate James Watson and later by Francis Collins, the project promised to revolutionize medical science. Mutations in genes can cause them to malfunction. Sequencing every gene held out the promise of being able to cure an untold number of diseases by correcting for faulty genes. There was added drama when government scientists began racing against a private company, Celera, to cross the finish line first. At stake was not only whether genes would be commercialized but also whether we would finally find effective cures for diseases such as cancer.[11]

With the scientific community fixated on DNA, whatever interest was still left in messenger RNA as a therapeutic nearly vanished. Making matters worse, RNA is difficult to work with in the laboratory. It requires an extreme level of cleanliness, beyond regular standards of sterilization. Once a lab was contaminated, it was forever useless. RNA is also delicate and fragile. Our bodies make it to last just long enough for ribosome to punch out proteins. But the biggest problem of all was one that would take Karikó years to discover and to solve.

Karikó was eager to continue her research of mRNA, and Barnathan gave her the green light. But the funding nightmare continued. Karikó was convinced that mRNA could be used as a form of therapy—inducing the body to make proteins. Barnathan agreed and offered whatever help he could. As a heart surgeon, he performed bypass surgeries and would give blood vessels left over to Karikó so she could use them in experiments. Karikó and Barnathan wanted to make a protein called a urokinase receptor. "Most people laughed at us," Barnathan said years later.[12] One day Karikó put mRNA on a blood vessel, and 20 minutes later she could see the RNA producing proteins. It happened so quickly. This was a big moment. "I felt like a god," Karikó recalled.[13]

Barnathan and Karikó started imagining all kinds of possibilities for mRNA. They even wondered whether they could use mRNA to make human cells live longer.[14] Barnathan tried to entice venture capitalists to invest in using mRNA as a new way to make drugs. But no deal ever went through. "They initially promised to give us money, but then they never returned my phone calls," Karikó recalls.[15]

Not long after that, Barnathan left Penn. Once again, Karikó's future was thrown into doubt. Given her inability to bring in grant money, Karikó was in a precarious situation. She had a boss who supported her but then landed a job at Harvard. The new boss gave Karikó a terrible choice. She could leave the university or accept a demotion.[16] Karikó had little choice. Humiliated, she accepted the demotion.

"Retrospectively she was maligned and not respected, but you know at the time, she was just like a fish out of water," said David Langer, a medical student turned resident who worked alongside Karikó for 11 years. "No one knew what to do with her."

"I think part of the problem is not that Penn did anything wrong. It's just that science is as much about politics and money" as about science, Langer said. "All the people that screwed Kate, they were doing it because of budget or because of space or because she hadn't done anything.

"She couldn't get a grant. She wasn't a particularly great grant writer. She was too busy doing the science. Grant writing is a downside of being a scientist . . . In an ideal world, you just get paid and you just do your work and you do whatever you want, but unfortunately that's not the way the real world works."

Karikó confesses that she actually liked writing grants. It made her imagine what she could do if she had the money. Getting rejected didn't hurt her feelings. Karikó's philosophy, harkening back to the Hans Selye book she read in high school, was not to take criticism personally. "I always listen, and I say, 'Maybe they're right. Maybe I'm not good enough. Maybe I didn't put enough data. I should do more.'"

Karikó was getting nudged out of the cardiology laboratory where she had worked for a decade, but Langer came to her defense. He was just a lowly resident, but he convinced the chairman of the neurosurgery department, Eugene Flamm, to hire Karikó to help him with some RNA research. Flamm had made neurosurgery an independent department.[17] Langer made the pitch to Flamm, telling him that Karikó was a great scientist. She could be valuable by helping Langer to explore new ways of treating cerebral vasospasm, a condition in which blood vessels narrow, reducing the flow of blood to the brain. It can kill brain cells, leading to transient ischemic attacks, or TIAs, or to full-blown strokes.

Although Langer was sincerely interested in the science, he was also motivated by his friendship with Karikó. He knew that if she were forced out of Penn, she would lose a huge tuition discount for her daughter, Susan, who hadn't graduated high school yet. Flamm bought Langer's pitch and hired Karikó.

Despite his deep admiration for Karikó, Langer is also defensive of those at Penn who couldn't figure out what to do with her. "At the time, there was a huge bias against RNA primarily because it was thought to be so unstable in the lab."

The university issued a statement for this book, without directly addressing questions about her being demoted or forced to find another job. The statement said, "We are grateful for Dr. Karikó's important contributions both during her time at Penn—where she continues to hold an appointment as an adjunct professor—and in her present role at BioNTech."

Karikó now got her own salary, rather than poaching off somebody else's grant. Langer recalls it was a paltry salary and estimated it at $40,000 a year. But she also got her own laboratory and was able to continue her work with messenger RNA. Langer's idea was to try to save patients with constrictions of the blood vessels in the brain by getting cells to make nitric oxide, which increases the blood flowing in the vessels but has a half-life of two milliseconds. Langer had patients who only had perhaps a day or two to live, Karikó recalls. "So he thought that if we deliver messenger RNA coded for nitric oxide synthase, then the patient will be salvaged and saved."

Part of the challenge was just getting the mRNA inside the cells. They were using endothelial cells harvested from umbilical cords in the maternity ward. Endothelial cells help prevent blood from clotting and control blood flow in the vessels.[18] They would then grow the cells in culture in petri dishes. But they weren't having much success getting mRNA into the cells. Finally, Langer flew to Davis, California, and spent three days at the home of Robert Malone, an RNA researcher who seemed to be having success penetrating cells using nanoparticles. When Langer got back to Philadelphia, he and Karikó tried Malone's techniques, but it didn't really make a difference. After COVID-19 struck, Malone would claim that he was the scientist who invented the vaccine technology platform.[19] He also became a harsh critic of the vaccines. Langer contends that Malone merely played a small role in the development of RNA science. With his contributions alone, there never would have been vaccines. "He had nothing to do with any of this," Langer said. "He's nuts."

Karikó and Langer were able to measure nitric oxide production in cells when they tested the mRNA. But they were never able to complete the research. Both Langer and Flamm left Penn in 1998 to go to New York. The new chairman of neurosurgery, Sean Grady, who is still in that position, did not seem to know what to do with Karikó. "It was like, 'Okay, she's there. What should we do with her?'" Karikó recalls. Langer said Karikó's laboratory was moved to a dingy basement that he described as like a "dungeon." It was once again appeared to be an attempt to nudge her out, make her unhappy enough to leave. But she was used to this by now. And she hung on.

"I don't think people didn't respect her," Langer said. "It was just that what she was doing was very leading-edge science. But she didn't have the reputation. She was a woman in a man's world.

She's a Hungarian scientist . . . She was six feet tall and was never afraid to tell you you're wrong. Didn't give a shit about your ego. She would challenge you, and this was very off-putting."

Karikó clung stubbornly to her pursuit of mRNA at Penn despite the lack of grants. She would find articles in scientific journals and head to the Xerox machine at the medical library. At the photocopier, she would have chance encounters with another hoarder of articles, someone who worked in a different building on campus. He was shy and not much of a talker. But that didn't faze Karikó, who would strike up conversations with him. Small talk at the copy machine in 1998 would end up not only changing her life but would eventually lead to a breakthrough in mRNA science.

CHAPTER 4

The Collaboration

Anthony Fauci was sitting in his office at a research hospital in Bethesda when someone laid the June 5, 1981, edition of *Morbidity Mortality Weekly Report* on his desk.[1] *MMWR* is often called "the voice of the CDC." It's a thin newsletter that tracks the latest information on infectious diseases. Little did he know it was about to radically change the course of his career.[2] In it was a perplexing writeup of five young gay men in Los Angeles who contracted a rare infection known as pneumocystis pneumonia, or PCP. This pneumonia was almost exclusively seen in people with dramatically compromised immune systems. Fauci was struck that those five healthy gay men would all suffer from such a rare disease. Two had even died. The mere fact that they were all homosexual led to immediate speculation that they might have a sexually transmitted disease.[3] Fauci had seen cases of this rare pneumonia in cancer patients receiving chemotherapy. He did not know what to make of the article. He assumed it was just a fluke, nothing to worry about.[4] But then a month later, he picked up the next issue of *MMWR*, dated July 3, 1981. It led with a disturbing three-page article on 26 cases of a cancer called Kaposi's sarcoma among gay men in New York and California. Kaposi's creates bluish lesions on the skin and on the mucous membranes inside the mouth. It was so rare that there were no good statistics on how often it occurred, but the best estimate was two to six cases per 10 million Americans.[5] Ten of the men had the rare PCP pneumonia as well. "The thing that blew me away was that all of them were gay men," Fauci later said.[6]

"It was at that point that I fully realized and knew, and I remember even got goose pimples about it, saying, 'Oh my goodness. This is a new disease,'" Fauci would later recall. "I made a decision in the middle of the summer of 1981 that I was actually going to change the direction of my career and start bringing into the hospital and studying these unusual situations of gay men who had this strange disease."[7]

Fauci, then an intense 40-year-old clinician and researcher with close-cropped jet-black hair, had no idea what this new disease was. Yet, he decided to sacrifice the research he had been doing on suppressing overactive immune systems to treat diseases such as lupus and rheumatoid arthritis. That research that had been groundbreaking and offered lots of promise.[8] Yet Fauci now wanted to devote himself to studying what he imagined would be a baffling and frightening new disease. "In a well-meaning way, my mentors, the people who had cultivated me in science and academics, thought I was being foolish to throw away a very promising career in one area of medical research to go after something that they thought was going to disappear."[9] He treated patients suffering from this still unnamed ailment in the hospital while his laboratory began doing basic research. Although they suspected it was a virus that was sexually transmitted, they also found cases in IV drug users and hemophiliacs, suggesting it was also transmitted through blood.[10]

A year later, Fauci would publish a brief editorial in the *Annals of Internal Medicine* about this still unnamed disease. By then, there were 290 recognized cases. But Fauci was concerned that this troubling new illness was not getting the attention it deserved. His words had an air of frustration and took up the cause for funneling resources to a disease that many dismissed as a gay illness. "Clearly, this extremely important public health problem deserves intensive investigation . . . Important information of scientific interest may ultimately result from study of this syndrome . . . However, the immediate goal that must be recognized and vigorously pursued is the designation of resources and energy to solving of the mystery behind the extraordinary disease, which currently seems to selectively affect a particular segment of our society. The population that currently is affected deserves this effort. Furthermore, because we do not know the cause of this syndrome, any assumption that this syndrome will remain restricted to a particular segment of our society is truly an assumption without scientific basis."[11]

By mid-1982, the disease known as gay-related immunodeficiency, or GRID, would finally get a permanent name: acquired immune deficiency syndrome, better known as AIDS.[12] Although the disease was incredibly deadly, killing all but the luckiest, it was mostly ignored by the media. It was also ignored by President Ronald Reagan. White House press secretary Larry Speakes would for years crudely laugh off the few questions he got about AIDS, turning them into gay jokes. Remarkably, the White House press corps would break into laughter. Speakes got the first question on the syndrome on October 15, 1982, from journalist Lester Kinsolving. It was featured in the documentary *When AIDS Was Funny* by filmmaker Scott Calonico:

Lester Kinsolving: Does the president have any reaction to the announcement by the Centers for Disease Control in Atlanta that AIDS is now an epidemic in over 600 cases?

Larry Speakes: AIDS? I haven't got anything on it.

Lester Kinsolving: Over a third of them have died. It's known as "gay plague." [Press pool laughter.] No, it is. It's a pretty serious thing. One in every three people that get this have died. And I wonder if the president was aware of this.

Larry Speakes: I don't have it. [Press pool laughter.] Do you?

Lester Kinsolving: You don't have it? Well, I'm relieved to hear that, Larry! [Press pool laughter.]

Larry Speakes: Do you?

Lester Kinsolving: No, I don't.

Larry Speakes: You didn't answer my question. How do you know? [Press pool laughter.]

Lester Kinsolving: Does the president—in other words, the White House—look on this as a great joke?

Larry Speakes: No, I don't know anything about it, Lester.[13]

In 1983, Congress earmarked a paltry $12 million for HIV/AIDS research.[14]

Despite the lack of interest, progress was being made. In 1983, a French team isolated the virus, and a year later Robert Gallo at the National Cancer Institute proved that the virus caused AIDS.[15] Around the time Gallo made his discovery, Fauci took over leadership of the National Institute of Allergy and Infectious Diseases, a sister agency within the National Institutes of Health. He would hold that same job for decades, even through the COVID-19 pandemic. In this new role, Fauci kept up the pressure to devote resources to finding a treatment and vaccine for AIDS. By late November 1984, the CDC reported there were 6,993 known patients with AIDS.[16]

"There were people who were concerned, and I would say bordering on being angry with me, in that it was clear that I wanted to put more resources in this. Because even though it was still very early in the history of the pandemic, I wanted to get more government resources, I wanted more research. And it became clear to me that we needed to embrace the gay community, the activists, to get a better feel for what was going on in the trenches with them. And there was a lot of resentment towards me on that. Resentment on the part of the scientists because they thought I was going to divert resources away from other important areas of infectious diseases. And I was arguing, 'I don't want to divert resources. I want to get new resources. I want to argue before the Congress and before the president about why we needed more resources for this disease.' So that was that area of resentment."

Fauci had his own lab with about 130 scientists who devoted themselves to the study of AIDS. He had to fight for funding and had mixed success. He even met pushback from colleagues in infectious disease at the National Institutes of Health. "There were some of the classical infectious diseases people who quite frankly felt very offended at that and made their thoughts and feelings known about that, thinking I was overemphasizing a disease that was only affecting a few thousand people," Fauci would recall later. "They're no longer here."[17]

One of the first scientists to join Fauci in his effort to research AIDs in 1981 was Drew Weissman. Weissman stayed in Fauci's lab as a fellow through six years of this tumult. He was at the center of medical research to try to understand and tame this puzzling new virus. "We didn't know much. There were no drugs available. It was really wide-open space to learn," Weissman recalls. There was no vaccine program at the time. Weissman focused on studying the biological mechanisms that led to the disease.

Weissman's life lacks the drama of Karikó's. He grew up in historic Lexington, Massachusetts. Its claim to fame is that Paul Revere alerted the citizens there to the advance of British troops and shortly thereafter it became the site of the first battle of the Revolutionary War. In a profile done by a local newspaper, Weissman's younger sister described his childhood as boring—an obedient child whose idea of fun even in grade school was to hang out at the library.[18] Weissman worked at his father's engineering company for a while and thought about becoming an engineer. But once he started at nearby Brandeis University, he realized how much he loved research. He graduated with a biochemistry major and then got a master's degree before entering the Boston University School of Medicine. He added a doctorate in immunology, the science of the immune system.[19]

Weissman had a stellar resume and landed a job in 1987 as an assistant professor at the University of Pennsylvania, where he started his own lab. He wanted to continue his research on HIV, including taking up the search for an HIV vaccine. The virus proved a formidable foe. For one thing, the virus constantly mutates, even inside a person's body, always staying a step ahead of the immune system. For another, it hides among the genes of a cell, meaning that the immune system cannot even find it. But what surprised Fauci and his researchers the most was that unlike almost any other disease, including polio or the plague, our bodies simply don't mount an effective immune response. Any vaccine would have to do better than our own immune systems.[20]

When Weissman went to the medical library, he had chance encounters with Karikó at the photocopier in 1998. They were both pack rats, hoarding copies of scientific articles. So they would often have to wait for each other to finish copying. Yet, in many ways, they were opposites. Weissman is quiet and laconic. Karikó is a chatty extrovert, darting from one topic to another when she talks. When Karikó learned that Weissman was working on an HIV vaccine, she blurted out an offer to make him RNA. It was the sort of thing for which she was known, so she wasn't sure whether Weissman was truly surprised.

"Certain people I already gave RNA. Whatever they needed. Most of them probably just put it in the freezer," Karikó recalls. "You know, [thinking] this crazy lady pushing this RNA stuff."

Weissman was sincerely interested. He wanted to research the possibility of using dendritic cells to create an HIV vaccine. Dendritic cells act as sentinels against foreign invaders. They live in small numbers on exposed tissues, such as on our skin or in our noses, lungs, stomachs, and intestines. Once they detect an invader, called an antigen, they analyze it and then dash off to other parts of the body to teach T cells how to attack.[21]

Dendritic cells were discovered by a young postdoc at New York's Rockefeller University in 1973 as he was trying to understand how white blood cells respond to invading microorganisms and to tumor cells. Ralph Steinman's discovery seemed to offer a new approach to vaccines. In 2011, Steinman won the Nobel Prize in Physiology or Medicine for the discovery.[22]

Very few scientists were interested in mRNA when Weissman and Karikó began their collaboration. In the 1990s, a few scientists tried to use RNA as a therapeutic but couldn't get it to work. Some tried to use it in vaccines for cancers and HIV, but they were not very good. Scientists would take cells from an animal, turn them into dendritic cells, insert the RNA in the cells, and then put the cells back into the animal. This led to problems. Scientists gave up.

Biotech companies were having more success with another offshoot of DNA science: recombinant proteins. In the 1970s, scientists learned how to insert genes into the cells of mammals or bacteria to produce proteins. Those proteins could be used to replace missing or defective proteins in humans. Genentech, partnering with Eli Lilly, became the first biotech company to produce insulin this way as a therapeutic.[23]

In 2018, an estimated 13% of the drug market's revenues came from recombinant protein drugs. That share continues to grow.[24] But the process is exceedingly complex, time consuming, and expensive. Proteins are grown in large vats called bioreactors, and the process involves thousands of steps. Recombinant proteins also cannot be used to replace proteins that remain inside cells, which is most proteins.

Weissman and Karikó started working together on the concept of using mRNA to replace drugs. Instead of making proteins in vats, they would code mRNA to make the proteins inside a human body. She was the expert on RNA. He was the immunologist who understand the immune system. "Kati Karikó and I started working on RNA essentially as a protein delivery system. We saw it as a transient gene therapy. And the idea was that it was easier, cheaper and probably better than trying to deliver a protein."

As simple as it sounds, if they succeeded, it would be difficult to overstate how revolutionary their breakthrough would be. It would be like hacking into the body's genetic process and recoding it to help it fight disease. Instead of making drugs at a factory, mRNA would allow your body itself to make the drugs internally. If you had a faulty gene making you sick, mRNA could come to the rescue, making the protein your body is failing to produce. It offered a different, seemingly simpler approach to gene therapy than trying to reprogram our DNA. The prospects are endless. There are 7,000 rare diseases that mRNA could tackle, not to mention cancer. And the easiest application of all is vaccines. "Vaccines are called low-hanging fruit. So what that means is that compared to developing the drug, they're relatively easy," Weissman explains.

As they were collaborating, Karikó's future was once again in doubt. "She kept writing grants. She kept writing papers, and nobody would give her money," Weissman recalls. What's more, David Langer had left before they could finish their research together. By contrast, Weissman came to Penn highly prized. Weissman soon turned out to be a rainmaker. In 1998, he received a $309,000 grant from NIH's National Heart, Lung, and Blood Institute to study antigens and HIV replication.[25] Although he couldn't use that grant to study mRNA, he could dip into a pool of cash the University of Pennsylvania gives new scientists until they have time to bring in their own grants. This is a common practice at research universities. Weissman used the startup money to cover Karikó's salary and the cost of her experiments. In 1999, Weissman received another $238,000 from the National Institute of Allergy and Infectious Diseases, run by his former boss Fauci. This time, he could spend the government money to pursue his interest in mRNA. And in 2002, Weissman was able to get a small business grant. Weissman, who appreciated the potential of mRNA, came to Karikó's rescue. It wasn't charity, however. He needed her expertise in RNA science.

Within a year or so of working together, they made their first breakthrough discovery. Karikó would give him mRNA that she had made with her own hands, not ordered out of a laboratory. If you put the mRNA directly on cells, they will just eat it up. So you have to coat it with something. Weissman would mix the mRNA with a fat bubble. He would then use a pipette, akin to an eye dropper, to place the mixture onto cells in a petri dish. When Weissman put mRNA on dendritic cells, he looked at them under a microscope. The cells had dramatically changed shape and size. It just so happened that dendritic cells give obvious clues when they are inflamed. The cells were rejecting the mRNA. We all have an adaptive immune system that fights off foreign invaders. But cells themselves have an innate immune response as well. The cells recognized the mRNA as a foreign invader and attacked it. That's what caused the inflammation. Weissman and Karikó realized that was the reason no one could get mRNA to make much protein in a cell. The cells were doing battle with the mRNA. As simple as it sounds, this was a crucial discovery that would forever change RNA research.

Weissman wasn't too concerned about the immune reaction. He wanted to make a vaccine and actually needed to create an immune reaction for the vaccine to be effective. Some vaccines today are made of proteins from a virus. The vaccine also had a chemical known as an adjuvant to help activate the immune response.[26] The point of a vaccine is to make the immune system create antibodies that will be ready to fight if the person were ever infected with the real virus. Antibodies are an immune response. So to Weissman, the innate immune reaction seemed like a good thing when it came to making a vaccine. "This is the perfect vaccine," he told Karikó. He wrote a paper on his discovery that was published in the *Journal of Immunology* in 2000.[27]

Karikó, on the other hand, was devastated. She was trying to use mRNA as a drug. If the cells attacked the mRNA, it would not make the protein she wanted. Her mission now became finding a way to evade the cells' immune response. It was work that would take years.

The duo had the same immune reaction during experiments in mice. They would give the rodents mRNA, but the mice wouldn't produce much protein. Worse, they would get sick. They hunched up. They stopped eating, stopped moving. Sometimes they died. Weissman and Karikó set out to figure out why the animals were acutely ill. Again, they realized that RNA was causing severe inflammation. So they started their quest to find a way to prevent RNA from causing a disastrous immune reaction.

Karikó began looking for ways to modify RNA.

Let's pause to talk a little bit more about RNA. To simplify, RNA is made up of four nucleosides, three of which also exist in DNA: adenosine, cytidine, and guanosine. It also has one nucleoside unique to RNA: uridine. Here is all you need to remember: These components of RNA are commonly known by their first letter: A, C, G, and U. Three of these letters in sequence will make one of 20 possible amino acids. A chain of amino acids will make one of more than 20,000 possible proteins in humans. That is why RNA is often called the software of life. Just as different combinations of ones and zeros make everything digital possible, different sequences of these four letters make life possible.

There are several types of RNA. The star of the show is messenger RNA. It takes the genetic code from the DNA and carries it to the ribosome to make proteins. But there is also transfer RNA, a cloverleaf-shaped molecule that helps transfer the mRNA's code to the ribosome. Then there is ribosomal RNA, which helps to make the ribosomes. Karikó's task was to find out how to elude cells' innate immune response by experimenting with different types of RNA. She also tried chemically modifying nucleosides. The dual strategies would become important to their breakthrough discovery.

Weissman and Karikó did countless experiments to see if RNA always caused inflammation. Sometimes it did and sometimes it didn't. What they realized over time is that some forms of RNA cause a lot less inflammation. In biology class, students are taught that there are only four nucleosides. But that is an oversimplification. RNA in our bodies is much more complex. Much of it is naturally modified. While the textbooks tell us the nucleosides are A, C, G, and U, in real life they are more varied than that. That led to an important discovery.

"When we looked at it, what we realized was that RNAs that had the most nucleoside modification had the least inflammation," Weissman recalls.

What Karikó and Weissman discovered was that when the basic form of unmodified mRNA was introduced into a cell, it was seen as a foreign invader. It lacked the slight changes that are natural inside our cells. Their research showed them that to elude the cell's immune reaction, it would make sense to modify the RNA first. Rather than putting plain-vanilla mRNA in a cell, create mRNA that would more precisely mimic naturally occurring modified mRNA.

The internet was still new then, so Karikó delved into scientific journals at the library. She was in a meeting one day with a Hungarian scientist, Tamás Kiss, from Toulouse, France. He had modified ribosomal RNA with a nucleoside called pseudouridine. It is closely related to the nucleoside coded as U for uridine. Karikó was staying at the same dormitory with the scientist and asked him if he could send her an enzyme that would allow her to modify RNA. "He said, 'Oh, that's not so simple. You need many components. It cannot be done,'" Karikó recalls. So she tried another old friend from Hungary, who told her she could buy a chemical from a laboratory to modify RNA.

Karikó would make textbook basic RNA, as researchers had been doing for years. But then she would use enzymes that would interact with the nucleoside to modify them. Scientists can modify RNA using synthetic nucleosides. But Karikó decided to avoid those. She only wanted natural nucleosides, and for good reason. Karikó recalled a disastrous clinical trial that had used an unnatural nucleoside to test a treatment for hepatitis. On June 30, 1993, the National Institutes of Health abruptly halted the trial when a 60-year-old man was admitted into a hospital with liver failure. Despite a transplant, he died.[28] By September, five out of the 15 patients in the trial were dead, all with liver failure. Two others were saved by transplants.[29] The medication, fialuridine, was a nucleoside analogue, which resembles RNA nucleosides.[30] Hepatitis, like many viruses, is made up of RNA. Researchers were using FIAU to try to prevent the virus from replicating. Although it had happened years earlier, Karikó made a mental note not to use synthetic nucleosides to modify mRNA.

Karikó criticizes the way scientists are trained in America. They become so highly specialized that they can lack basic knowledge. In Hungary, Karikó wasn't always able to just order the products she needed. Often, she had to make them with her own hands. For example, she made the chemicals to analyze fish fat when she was working at the fishery in Hungary. At Penn, she made her own RNA. As a result, she had broad and deep expertise in RNA. "You have to know these things: How the RNA is made. How it is translated. What is the degrading enzyme. What inhibits the degrading enzyme. How it comes in. How it goes out. As a Hungarian, we learn everything," she said.

Just as Karikó and Weissman began delving into using RNA as a form of gene therapy, there was a tragedy on campus that would dominate national news. At Penn's Institute for Human Gene Therapy, a highly acclaimed scientist, James Wilson, was researching a rare metabolic disorder called ornithine transcarbamylase (OTC) deficiency. This condition is caused by a mutation in the OTC gene on the X chromosome. A person with this disorder doesn't produce enough OTC enzymes to flush out ammonia, a toxin produced by the breakdown of protein. Half of babies born with OTC deficiency die.

In Tuscon, Arizona, 17-year-old Jesse Gelsinger read about Wilson's research. He had a milder version of the disorder that he could control with a low-protein diet and about 50 pills a day. Still, one day, after not taking his medicine, Gelsinger ended up writhing on his couch at home, throwing up nonstop. He had to be induced into a coma to recover.[31]

Gelsinger understood that gene therapy was unlikely to cure him of his disease. But he wanted to do his part for science. He especially wanted to help newborns with the condition. Gelsginer told friends that if the worst were to happen, "I die, and it's for the babies."

He enrolled in Wilson's clinical trial as soon as he became eligible at age 18 in 1999.

By then, Wilson's clinical trial had been underway for almost two and a half years. Gelsinger was the eighteenth person to be dosed with the vector, the second patient in the sixth cohort. Some of the subjects had reported fevers or flu-like symptoms, but nothing serious.

Gelsinger received the adenovirus vector containing the engineered genes on September 13, 1999. His immune system reacted fiercely. He developed jaundice and blood-clotting problems. His kidneys and lungs failed. Gelsinger was declared brain dead on September 17—98 hours after his first vector.[32]

Researchers never confirmed why Gelsinger suffered such a toxic reaction to the gene therapy. But his death set off a barrage of negative headlines, investigations, and recrimination. Gelsinger's father, Paul, initially stood by Wilson and the University of Pennsylvania. But he later filed a lawsuit, alleging that his family was not informed that monkeys had died in previous experiments and that Wilson and the school had big financial stakes in the outcome of the adenovirus trial. The lawsuit was settled out of court with no admission of wrongdoing.[33]

Wilson lost his titles and was barred from any more clinical trials until 2010. The Institute for Human Gene Therapy was shut down. It sent a strong message to the scientific community that gene therapy was a bigger challenge than we thought. The promise of the Human Genome Project that was still under way was that a deep understanding of DNA would revolutionize medicine. Drugs would be replaced with gene therapy. But Gelsinger's death was a major setback.

Using RNA doesn't carry the same risks. RNA doesn't alter our genes; it only carries out their function and then quickly dissipates. The young man's death might have renewed interest in mRNA research as an alternative form of gene therapy. But that did not happen. Weissman and Karikó remained within a tiny universe of scientists paying any attention to RNA.

The experiments Karikó and Weissman were doing required intimate knowledge of mRNA. In our cells, DNA makes RNA with the help of the enzyme polymerase. The enzyme makes a copy of the DNA that then makes the RNA. Karikó and Weissman would do the same thing in a test tube. They would take DNA and add an enzyme called RNA polymerase. The enzyme copies the DNA and makes RNA from it. At one point, Karikó spent $5,000 to order synthetic RNA from a lab. But it turned out to be a waste of money. So she went back to making her own RNA.

Karikó began to realize that the nucleoside uridine was critical in causing inflammation of the cells. So she looked for ways to modify it. She might have used messenger RNA, but one day she had basically run out of it. So she tested transfer RNA instead. She figured that RNA is RNA. One day she tried 10 different modifications. Five of them worked. During the experiment, she went back to her office to look something up in a book. She was so excited that she ran back to the laboratory, thinking later that she must have looked silly. But when she replaced uridine with pseudouridine, the one suggested by the Hungarian scientist, she made an amazing discovery. Not only did pseudouridine evade the immune response, but it even made more protein. "My God," Karikó thought as she realized the results. "Now I can make my mRNA therapeutics."

She went to tell Weissman. As they both realized what they had just done, they didn't quite believe it. They quickly tried it again and again. And they got the same results. "We looked at each other and said, 'This is it!'" Weissman recalls. "We figured out that this is going to make RNA therapeutics the treatment of the future."

That was in 2004. In hindsight, the answer seemed so obvious. "How is it that people didn't see that uridine is the answer?"

Karikó wondered. In fact, she remembers reading a paper from 1963 that came close to solving the riddle. Now Karikó and Weissman wrote their own paper. They knew that they were about to revolutionize mRNA science as well as medical science. So they went for the most prestigious journal possible: *Nature.* But the journal rejected the paper instantly. It came back with a note that "whatever we were describing is just 'incremental,'" Karikó recalls. Her English is still a work in progress, so she had to dig up a dictionary and look up the word *incremental.* After reading the definition, she thought, "Oh, they weren't impressed."

They then submitted it to the journal *Immunity.* After revisions based on comments from peer reviewers, it was finally published on August 23, 2005. The headline was imposing: "Suppression of RNA Recognition by Toll-like Receptors: The Impact of Nucleoside Modification and the Evolutionary Origin of RNA." The article itself is impenetrable to a lay audience and likely to most biologists. The first sentence read: "DNA and RNA stimulate the mammalian innate immune system through activation of Toll-like receptors (TLRs). DNA containing methylated CpG motifs, however, is not stimulatory."[34]

Their discovery opened a whole new field of therapeutics, one that by then the scientific community had thought about a few times but rejected. And yet, their discovery garnered absolutely no press coverage. The scientific community itself ignored the article. "Really very few people cared or read it or were interested in it," Weissman said.

The attitude within the scientific community was that people had already looked at RNA, and there was nothing to see. "The trends had moved on. Nobody cared about RNA," said Chad Cowan, an assistant professor at the Harvard Stem Cell Institute. "Unfortunately, Karikó and Weissman were coming in at the tail end of interest in RNA biology." What's more, their paper was highly technical and lacked any talk of the incredible potential of RNA as a replacement for drugs. "They're bad communicators," Cowan said. "They don't exactly tout the [potential]. They don't explain, 'Here's what you could do with it.'"

The reaction was deflating, but Karikó and Weissman remained certain that their discovery had enormous potential. They worked together to file a 55-page patent application in August 2006. The first line read: "A method for inducing a mammalian cell to produce a protein of interest comprising: contacting said mammalian cell with in vitro-synthesized modified RNA encoding a protein of interest, wherein said in vitro-synthesized modified RNA comprises the modified nucleoside pseudouridine."[35]

After they filed for the patent, Karikó and Weissman tried to negotiate with the technology transfer office at the University of Pennsylvania to license the rights to their own invention. The rules at Penn posed a challenge. Karikó and Weissman were not allowed to negotiate on their own. They had to hire a lawyer. But their lawyer was unable to reach an agreement for a reasonable price. "What Penn was asking for was just unreasonable and wouldn't allow a small company to survive," Weissman said. Karikó and Weissman created a new startup company, RNARx LLC, and applied for a research and development grant from the federal Small Business Administration. The money would be used to show that they could produce the protein erythropoietin in animals using mRNA. EPO, as it is better known, is infamously known as one athletes use to stimulate red blood cell production to give themselves an unfair advantage during competition. Disgraced cyclist Lance Armstrong admitted in an interview with Oprah Winfrey that he injected EPO to help him win seven Tour de France titles that were later stripped from him.[36] But it has a legitimate use as well. EPO can be used to treat anemia. The grant was delayed because of a budget freeze during President George W. Bush's administration. In 1997, SBA awarded them $97,000. It promised to give them another $799,000 if they could prove it. But the University of Pennsylvania raised questions about whether the grant posed a conflict of interest for Karikó and Weissman. So they weren't able to use the money until that issue got resolved. Two years later, they received the rest of the grant.[37]

Over the years, they would talk to venture capitalists about investing in the company, but no one was interested. "People say that I was not a good communicator. I couldn't convince them," Karikó recalls. Yet Karikó notes that others tried and failed too. Two MBA students working on a project had gone to the technology transfer office and selected the RNA project to write up a business plan for a competition. But when the MBA students presented their plan, they couldn't convince the judges either.

Karikó knew that pseudouridine was naturally found in transfer RNA. In fact, it is the fifth most abundant nucleoside in different types of RNA in the human body. But she didn't know that pseudouridine was naturally found in messenger RNA. That discovery wouldn't be made until 2014. That's how mysterious RNA remained.

They kept doing experiments. They didn't know yet whether modified messenger RNA would cause an immune response. Karikó tried to produce different proteins in mRNA, starting with a protein called luciferase. It's what gives lightning bugs the amazing ability to glow in the dark. The protein emits a cold light when it mixes with a substrate. Karikó used a device called a luminometer to measure the light emitted from the chemical reaction. Karikó would put the mRNA on rabbit cells in a dish.

She was shocked to discover that when the mRNA was modified with pseudouridine, the cells in the petri dish would glow much more than when the mRNA was not modified. In fact, the modified mRNA produced twice the amount of protein.[38]

Next, Karikó tried the experiment on live mice. She gave eight-week-old mice an anesthetic and placed them on a heating pad. She then injected into the vein in their tails different doses of modified and unmodified mRNA coded to produce the same glowing protein. The mice were then killed and dissected. Their spleens were used to detect the protein. Again, she was surprised by the results. The modified mRNA eluded the immune reaction, but beyond that it was more stable and produce far more protein. Once again, they had made a major discovery that had huge implications.

Karikó wrote another article. Despite the significance of their finding, Karikó submitted it in February 2008 to the editors of a lesser-known journal, *Molecular Therapy*, the journal of the American Society of Gene Therapy. Although gene therapy was a hot topic, the journal lacked the sex appeal of the more elite journals. This time, though, Karikó and her co-authors made more of an attempt to explain the significance of the research. The paper was published September 16, 2008. In the discussion section near the end of the paper, Karikó framed the research as helping to create a new form of gene therapy using mRNA. So far, she noted, efforts to use mRNA had been limited to vaccines. But her findings made it possible to use mRNA in a whole host of possible therapies. She wrote, "These collective findings are important steps in developing the therapeutic potential of mRNA, such as using modified mRNA as an alternative to conventional vaccination and as a means for expressing clinically beneficial proteins in vivo safely and effectively."[39]

Despite the promise, the article once again attracted no attention—not a single news article. The scientific community remained unaware of it. Yet, Weissman and Karikó were still convinced that their discoveries could change medical science. As talks with their own university over rights to their own patent broke down, others showed interest in it. One was a newly formed company called Cellscript LLC, a subsidiary of Epicentre Biotechnologies, based in Madison, Wisconsin. Gary Dahl, the head of Epicentre, said Cellscript was formed to explore using mRNA both for clinical research and as a replacement for drugs. Dahl had read Karikó and Weissman's article and realized its potential. So he reached out to them to see if they had filed a patent and, if so, if he could license it. Karikó told Dahl that she and Weissman were trying to license the patent themselves for their own company. Without giving much detail, Karikó believes that Dahl paid $300,000 for exclusive rights to the patent. That was much less than what Penn was asking from Karikó and Weissman, but both said that those conversations are confidential. The university declined to discuss the details of the license.

Meanwhile, another company was also interested. CureVac, which was trying to develop vaccines with unmodified mRNA, talked to Penn about licensing the rights. Karikó said the company came very close to making a deal but decided against it, perhaps because it didn't want to pay $300,000. She believed it was a mistake. Without modifying its mRNA, CureVac was trying other techniques to get around the immune response and the lack of protein production. In Karikó's eyes, it was a futile approach.

On February 5, 2010, Karikó and Weissman were sent a letter from the associate vice provost for research, Michael Cleare, who oversaw selling the rights to intellectual property. It was deflating news. Gary Dahl was back, and he wanted to buy the rights to their discovery. "As we have been unable to reach agreement with RNARx, we have determined to speak with EPICENTRE Technologies Corporation concerning a possible license to the RNA technologies you have developed at Penn." Cleare said that if talks with Epicentre fell apart and Penn licensed only a portion of the rights to them, RNARx would be able to return to the negotiating table. But now, he needed them to agree that Penn could sell the rights to their patent. Two weeks later, both Karikó and Weissman signed. Penn, in a highly unusual move, sold all the rights to Epicentre. Weissman and Karikó were puzzled. "They usually never give full rights unless it's a pharmaceutical company that's done clinical trials," Weissman said. "I don't understand why they did that."

Karikó remains bitter about it. She accuses the university of selling the rights at a huge discount. When asked why Penn sold the exclusive rights, Karikó didn't hold back. "Because they are assholes," she said. Weissman believes the university gave away a major discovery without realizing that it had made a big mistake. Does the school know it now? "I would hope that they realize it. Every time they talk to Kati, she tells them how much they screwed up." Karikó goes so far as to say she has forgiven nearly everyone who dismissed her talents. She even forgives the boss who threatened to deport her. But there is one person she says she'll never forgive: the guy who sold the rights to her patent to someone else.

Even had Karikó and Weissman gotten the rights to their invention, Karikó's old colleague, David Langer, wonders whether they could have successfully commercialized the discovery. What's more, he thinks it would have been difficult for people at Penn, such as Karikó's boss, Sean Grady, the chairman of neurosurgery, to know the significance of their findings. "How can you blame anybody?" Langer asked. "Grady's a neurosurgeon. How the fuck is he supposed to know? Hell, RNA biologists didn't even know, and it was there in plain sight."

Still, Karikó wonders to this day what would have happened had CureVac licensed the patent. They had already spent a decade trying to make vaccines from mRNA without success. The answer was right before them. Said Karikó, "If they, with a little money, would invest and pay for it, there would not be a Moderna. There would only be CureVac."

CHAPTER 5

Scientific Sabotage

Twenty-six-year-old Facebook creator Mark Zuckerberg graced *Time*'s cover as Person of the Year in 2010. Inside, the magazine featured 50-odd "People Who Mattered." The list was full of obvious and instantly recognizable names, from Barack Obama and Pope Benedict to LeBron James and Lady Gaga. But there was at least one unknown: Derrick Rossi. The magazine identified him as a researcher whose discoveries "could help move stem cell–based treatments for diseases such as diabetes and Parkinson's more quickly from the lab to the clinic."[1]

Just four months later, in April, *Time* once again featured Rossi on a list of the world's most influential people of 2011. He was one of only 13 people to make both issues. Again, the magazine lauded Rossi for discoveries that could one day cure intractable diseases as well as treat spinal-cord injuries.[2]

Time made no mention—frankly, it was not well known—that the 45-year-old assistant professor at Harvard Medical School had just founded a biotech company in Cambridge, Massachusetts. He called it Moderna. Even then, there were company insiders who believed that Moderna would one day revolutionize medical science. No one could have guessed, however, that 10 years later Moderna would become the National Institutes of Health's choice to try to halt the worst pandemic in a century.

Shocking medical discoveries are exceedingly rare. Most advances in science are incremental. Rossi's breakthrough trumpeted by *Time* was amplified because it advanced a truly startling discovery. On June 30, 2006, Shinya Yamanaka, a doctor and researcher at Japan's Kyoto University, gave a morning presentation at a conference in Toronto that stunned a roomful of scientists.[3] Yamanaka, frequently consulting his notes, explained in his quiet voice that he had been able to take mature skin cells from mice and convert them back to embryonic stem cells. There was a hush in the room. Some could not believe what they were hearing. It was akin to biological time travel, like turning a chicken back into an egg.

In an embryo, stem cells are identical blank slates that divide and multiply. But within days, they will morph. Some will become liver cells. Others brain cells. Yamanaka was describing a discovery that might allow scientists to take cells from one part of the body, such as the skin, and convert them back to embryonic cells. Then, these cells could be coaxed into becoming an entirely different type of cell. Skin becomes bone marrow or maybe even a kidney. "From the point of view of moving biomedicine and regenerative medicine faster, this is about as big a deal as you could imagine," Irving Weissman, a leading stem cell biologist at Stanford University, told the *New York Times*.[4]

Yamanaka discovery of so-called induced pluripotent stem cells—IPS cells for short—had almost surrealist implications for future transplant patients, whose lives often depend on waiting for a perfect donor. Yamanaka was describing a method that might allow a leukemia patient to get a stem-cell transplant using his or her own cells—a perfect DNA match. No more worries about the body's immune system rejecting the transplant. Rather than waiting months for a new liver, someone with end-stage liver disease could instead have a new liver grown with cells containing their own DNA. Of course, these types of breakthroughs would require many more years, if not decades, of research.

It also had political implications. Using human embryos to do research became highly controversial dating back to the landmark *Roe v. Wade* decision, which legalized abortion. Stem cells are used to research diseases and to help find new treatments. This is done by studying different diseases with cells in a petri dish as opposed to doing the research on a person with the disease. Those embryonic cells typically come from fertilized eggs left over from in vitro fertilization. Even though the vast majority of these embryos will remain frozen in storage indefinitely, removing the stem cell kills them, destroying any chance of implanting the embryo into a woman and impregnating her. Critics consider this akin to abortion. In 2001, President George W. Bush signed an executive order restricting the use of federal funds on research relying on human embryonic stem cells other than those cell lines available before August 9, 2001—a totally arbitrary date.[5] But it forced scientists into studying other cell avenues, such as cloning. Yamanaka's work raised the prospect that you could create an embryonic stem cell that was not derived from a fertilized egg.

Rossi was in the room listening. He knew instantly that he was witnessing a rare historic moment. The science was so elegant, so awe inspiring. In the row just in front of him were two stem-cell heavyweights, Rudolf Jaenisch from MIT and Kevin Eggan of Harvard, who seemed transfixed by the presentation. Rarely do naturally skeptical scientists walk out after the presentation in awe. The unknown Japanese scientist became the buzz for months to follow. Just six years later, Yamanaka would take the stage in Oslo, Norway, to accept a Nobel Prize for his discovery.

But it was immediately obvious to Rossi that Yamanaka's method had drawbacks. For every cell Yamanaka could convert back to its embryonic state, he failed on more than 99 other cells. What's more, he used a virus as a tool for inserting DNA into a cell to reprogram it. Viruses are great at penetrating cells. Retroviruses can even penetrate the well-shielded nucleus of a cell, giving it direct access to our DNA coiled up inside. Using a virus to gain access to a cell is an everyday trick for biologists. They put DNA inside a retrovirus to do experiments. But it is a method fraught with problems in the real world of medicine. Retroviruses integrate into our DNA, tinkering with our genetic makeup. Like it or not, about 8% of our DNA has been reprogrammed by retroviruses from the distant past.[6] In Yamanaka's approach, he introduced four genes, known as Yamanaka factors, into the genome. Once they were inserted, they would always be part of the DNA. That can be a risky approach if anything goes wrong. In fact, gene therapy has been linked to cancer. It was obvious that another tool would be needed if Yamanaka's discovery were ever to be realized. This would ultimately create a race among academic laboratories to find an alternative.

At the time of Yamanaka's presentation, Rossi was working at Stanford University, where he was researching aging and working in the lab of Irving Weissman, a legend in stem-cell science. But he had grown up in Toronto, the son of working-class immigrants from Malta, a small island country south of Sicily. Rossi's father worked in an auto body shop. His mother owned a bakery, but she became weary of the 90-minute commute by bus and subway. She eventually gave up her bakery to start a day care in the neighborhood, where she charged people based on what they could pay.

Rossi stayed in Toronto for college and a master's degree. But then he decided to see the world. He enrolled at the University of Paris, where he says he spent too much time partying to get his doctorate. He went to Texas briefly before moving to Helsinki, where he got his PhD and did research that drew attention among academics. He fielded several offers, but in 2003, he landed as a postdoctoral fellow at Stanford doing stem-cell research. Four years later, he was hired by the Immune Disease Institute, a small 50-year-old nonprofit on the campus of Harvard Medical School on the Boston side of the Charles River. That also gave him a faculty appointment at Harvard Medical School and eventually made him an employee of Boston Children's Hospital, directly across the street from the school.

Friends describe Rossi as a bit of a rebel. "He was an iconoclast in the sense that he was doing things his way regardless," said Chad Cowan, a close friend who met Rossi when he was a student at the University of Texas Southwestern Medical Center. "He definitely always spoke his mind." Cowan described Rossi's look as a "cross between a rock star and a scientist." With his thick curly hair and goatee, Rossi resembled a young Robert Downey Jr. Rossi could be unpredictable. One time when Cowan hosted a big party at his home, Rossi brought a guest. When Cowan asked who the guest was, Rossi said, "Hey, you remember this guy. He's Ted. He's the homeless guy who lived in the van next to the laundromat we use."

When he arrived at Harvard, Rossi brought a colleague from Stanford with him who himself was a colorful character. Luigi Warren is an intense native of Great Britain. He grew up in a well-to-do family in Kent just southeast of London. His father was a prominent public servant who helped Nigeria gain its independence from Great Britain. Warren was fascinated at an early age with the works of science fiction authors, especially Robert Heinlein. He was glued to the TV when black-and-white images were broadcast live of Neil Armstrong taking humankind's first steps on the moon. Warren started his career as a skilled software engineer. He did some work for Sony Pictures Imageworks, a state-of-the-art visual effects and animation studio. He was the lead programmer for an action video game called Meat Puppet, set in a post-apocalyptic world. But it wasn't the kind of science he dreamed of doing as a child. So in midlife, Warren decided to start all over, getting a second bachelor's degree, this time in biology. He graduated from Columbia University summa cum laude in 2001 at the age of 40. He went directly to Cal Tech for his PhD, but by 2004, he began work at Stanford in a lab affiliated with Weissman's lab, where Rossi was a postdoc. Warren was the type of scientist who would get lost in his work and would never give up. Warren was thin, but a couple of friends recounted how at a Thanksgiving dinner and at a party he was able to eat far more than anyone else. One recalls Warren saying he would sometimes skip meals, a story Warren finds amusing but doesn't remember. Warren was a habitual chain smoker, going through as many as 40 cigarettes day. He was often found outside with other smokers. He and Rossi bonded over their passion for rock-and-roll music. Rossi loved David Bowie. Warren came from the town were Bowie lived in the 1960s. Warren had the look of a rock singer. He wore the same outfit all the time: black jeans, a T-shirt, and a jacket. "He looked cool," Chad Cowan said.

The two teamed up on a project and published a paper together on aging.[7] Rossi admired Warren's meticulous approach to projects, calling him a "brilliant technologist." But he also knew that Warren had an unusual passion. He maintained a blog on the internet with theories about the origins on the 9/11 attacks.

Despite evidence to the contrary, Warren was convinced that Saddam Hussein was the mastermind behind guiding passenger jets into the World Trade Center and the Pentagon. He also believed that Hussein was behind a series of anthrax-laced letters sent in the mail shortly after 9/11, killing five Americans and forcing 17 others to seek medical treatment. After a seven-year investigation, FBI director Robert Mueller held a press conference to announce that the Justice Department was just about to bring criminal charges against an anthrax researcher, Bruce Ivins. But just before charges could be filed, Ivins took his own life.[8] Warren never bought the findings of the FBI.

"Even today that whole question of the anthrax is almost unmentionable," Warren said. Warren thinks scientists are expected to accept conventional wisdom when it comes to matters outside their expertise. Yet he remains convinced that Hussein had the anthrax letters sent to deliver a clear message. Hussein was telling the US government not to meddle with him. Hussein was upset, according to Warren, because of several US-led attempts at assassinating Hussein and overthrowing his government in Iraq. Hussein's message, in Warren's mind, was, "I've got weapons of mass destruction and I'm in your country . . . I can bring the house down on you." Warren had blogged about this for years.

One day Warren came to Rossi in the lab at Stanford and asked if he would walk outside to have a talk. They went to a scenic courtyard outside the Beckman Center on the medical school campus. Warren asked Rossi if he had been getting any threatening phone calls or emails. Rossi shrugged, saying he hadn't. But Warren was convinced that he was under government surveillance, perhaps by military intelligence. He suspected they were interested because of his blog. He told Rossi that he had gone out for tacos one night and had bumped into a man who he thought was following him. Worried about his safety, Warren confided to Rossi that he had bought a gun. Warren's penchant for conspiracies would later threaten to derail the publication of the research paper that led to Moderna's creation.

Rossi was befuddled but put it out of his mind. He had been incredibly impressed with Warren's deep understanding of science. In that regard, he considered Warren to be brilliant. At some point, the two agreed that Warren would join Rossi in Boston. They were close enough that Rossi even picked up Warren at the airport when he arrived, in part to save him a few bucks.

Within months, Warren would dive into Yamanaka's work to find a better solution. The work started after the International Society for Stem Cell Research conference in Philadelphia in June 2008. The conference was packed with 2,500 attendees. Yamanaka was there again, and the buzz around his discovery was still reverberating. "There's certainly a bit of Shinya-mania," the ISSCR president told the prestigious scientific journal *Nature*.[9]

One of Warren's lab mates, Isabel Beerman, had invited a young Yale scientist she knew to have lunch with Warren and Rossi at the Hard Rock Cafe two blocks from the convention center. Rossi showed up late. The guest was Han Lee, who had done some work with messenger RNA. Lee said he began describing the benefit of using mRNA not only with stem cells but for therapeutic uses as well. Warren wanted to hear more. That night, at a Mexican restaurant on South Street in Philadelphia, Warren and Lee carried on a nerdy what-if conversation over margaritas. Rossi wasn't there. Warren was starting to wonder what would happen if instead of using a retrovirus to take a cell back in time, you tried using RNA. He had become comfortable making RNA in the lab at Stanford and had gotten over RNA's bad rap for being unstable. Lee recalls the conversation being lively, punctuated by

Warren's complaints to the waitress for bringing him the wrong size of margarita. Beerman was embarrassed and told Warren to leave the waitress alone. Lee thought it was funny. Eventually, the annoyed waitress let everyone at the table have their margaritas for free. The trio went out for more drinks after dinner.

We all know DNA is the secret to life. DNA stands for deoxyribonucleic acid. It is a double-stranded helix embedded in the nucleus of cells. DNA determines whether we are redheads or brunettes, whether we are tall or short. But more importantly, second by second, DNA transmits the genetic signals that keeps our bodies functioning. Need insulin to keep our blood sugar in check? DNA sends the signal to the pancreas to produce the insulin. Need an antibody to fight off a virus? DNA takes care of that. At any moment, this is happening within the 30 trillion cells in our bodies.[10]

DNA sends those signals by making RNA, short for ribonucleic acid. It is RNA that delivers DNA's message to the cell to make insulin or antibodies. Single-stranded RNA escapes a cell's nucleus and makes its way to ribosome, a sort of biological 3D printer. Ribosomes read the code on the RNA and produce amino acids that link together to become proteins. Genetic scientists have a saying coined by Francis Crick that is the central dogma of molecular biology: DNA makes RNA makes proteins makes life. Warren had an idea. What if you used RNA to tell cells to introduce the Yamanaka factors that would revert them to their embryonic state? He could not wait to get back to Boston to try out his theory.

Back in his lab, Warren wrote in his notes dated June 27, 2008, that he planned to start experimenting on IPS cells. Warren ordered some inexpensive ingredients he needed to begin learning how to make synthetic messenger RNA.[11] Once he had built up a bit of confidence a few weeks later, he took his idea to Rossi. Again, he filed an update. "Discussed IPS project with Derrick and got okay to proceed, as long as expenditures remain modest."[12] It was a bit out of character for Rossi, who usually didn't care about expenses. He would eventually pay for the research with a grant from the National Institutes of Health.

Warren started by trying to prove a simple point. If you put mRNA onto mouse cells in a petri dish, would it produce protein? He chose an easy protein to test—one that glows fluorescent green. Found naturally in jellyfish, it is called—quite unimaginatively—green fluorescent protein. Warren put the mRNA on the cells. After a while, he peered into the microscope and felt his heart beating faster. He rushed down the corridor to fetch Rossi. Rossi stared into the microscope and could not believe what he saw. The cells were glowing green. Cells from mice had produced a protein that comes from jellyfish. The mRNA worked.

That day in the summer of 2008, Warren and Rossi thought in their giddy excitement that they had found the solution. After that, Warren recalled that Rossi no longer cared how much it cost. Despite their excitement, the truth was this simple experiment was not particularly innovative. In fact, a German startup called CureVac had already been toiling away for years trying to make cancer treatments and vaccines using mRNA. That meant that CureVac was routinely using mRNA to make proteins.[13] Even so, the scientific community was all but ignoring mRNA at the time. So the basic act of expressing a protein using mRNA felt new and exciting. Granted, out of perhaps a million cells in a dish, only a thousand or so glowed. But Warren and Rossi were still blown away with the prospect. Now the task was to keep them glowing for days, since it takes that long for a cell to convert back to its embryonic state using Yamanaka's techniques—two weeks in mice cells, to be exact. RNA is notoriously unstable and temporal. To keep the proteins glowing, Warren was going to have to add more mRNA each day. But each day, the cells grew dimmer and dimmer and eventually stopped making the protein. One day, he had nothing but a plate of dead cells. His excitement flatlined.

The hypothesis was so simple, so obvious. Yet now, it finally dawned on Warren and Rossi why so few scientists were talking about RNA. Rossi imagined that hundreds if not thousands of laboratories had done the same experiment, and when the cells died, the scientists simply gave up. No one even wasted the time to publish about the failure of mRNA to produce proteins. Everybody assumed it was just a dead end. But Rossi and Warren were not ready to give up.

In October 2008, Warren decided to talk to Sun Hur, another scientist who ran a lab at the Immune Disease Institute. She was aware of work of Katalin Karikó and Drew Weissman. The mouse cells in a dish saw the RNA as a threat and attacked it. Everyone knows that our complex immune system responds to foreign invaders. But individual cells themselves have an innate response to anything suspicious as well. In this case, the cells were so overwhelmed that they died. Hur had several suggestions, but the key one was that Warren should consult Weissman and Karikó's 2008 paper in the scientific journal *Molecular Therapy* on how to modify RNA to get around the cell's immune reaction. Rossi and Warren gave modified RNA a nickname: mod RNA.

Armed with this method, Warren tried again. He would put mRNA on the cells every 12 hours. The first try did not work that well. And Warren began to suspect the problem was the mouse cells he was using. After going down several blind alleys, Warren began using human skin cells instead. He started again with the green fluorescent protein and was able to keep the cells glowing for days. Keeping the cells green for days—a scientific magic trick of sorts—was the moment when Rossi realized that by employing Weissman's and Karikó's methods they were about to do something groundbreaking.[14]

Warren filed an update on November 2, 2008: "Looks like a winner here."[15]

There was still a lot to do. Warren had to convert the cells to their embryonic state, which involved using the mRNA to insert four genes into the cells. This would take another year to accomplish. But in the meantime, Rossi and Warren met with Ryan Dietz, who helps scientists patent inventions and sells the rights to discoveries. They brainstormed about how to make money off the project. Rossi was ready to start a company based on mRNA science. He had the same epiphany as Weissman and Karikó. Mod RNA was a game changer. Although Rossi's work started with creating pluripotent stem cells in a petri dish, he realized that using modified RNA to transplant organs was a distant dream and potentially not commercially viable. But there were other potential uses of induced pluripotent stem (IPS) cells in the near term, such as using IPS cells to study disease. Researchers would be able to create a diseased cell from a patient on which to test drug therapies. Rossi says he understood from the start, just like Weissman and Karikó, that the real potential of modified RNA was its ability to turn our bodies into drug-making machines. In a series of meetings in early 2009, Rossi discussed the idea of using mRNA as a therapeutic with Dietz, according to Dietz's notes at the time.

The experiment was not yet successful, however. Warren was having technical problems with the RNA that he was making. So he consulted with a manufacturing company called Epicentre, based in Madison, Wisconsin. Epicentre had two years earlier created a wholly owned subsidiary, Cellscript. The head of the company, Gary Dahl, said its goal was to use mRNA to express proteins in cells to be used in clinical research and therapeutics. Warren could tell that the people he was talking to had real expertise with mRNA. But they didn't seem to be aware of the work published by Karikó and Weissman. As a favor, Warren mentioned to them that they might want to look into it. Not long after that, Dahl contacted Weissman and Karikó to see if there was a patent and if they could sublicense it. Yes, there was a patent on file. But no, they were not interested. They were trying to start their own company. In academia, the university owns any patents filed by faculty as part of their work. The talks between Penn and its two researchers broke down eventually. At that point, Cellsript swooped in and talked dealmakers at the Ivy League school into selling it the exclusive rights to the patent. Cellscript itself never followed through on any effort to pursue mRNA as a therapeutic. Warren later suspected that Dahl could see that someone would soon start using mRNA as a drug. And to do that, they would have to sublicense the patent from him. Perhaps it was not just about pursuing science but more about leveraging intellectual property—a pure paper transaction. Dahl had managed to get the exclusive license for a pittance. It would one day pay off big.

In the meantime, there was growing friction between Warren and Rossi. Warren kept hearing rumors that other labs were close to solving the retrovirus problem by reprogramming IPS cells with RNA. An acquaintance had gone to a lecture in Leipzig, Germany, and heard a researcher from the Fraunhofer Institute say that that she was already creating IPS cells using mRNA. The acquaintance sent an urgent email to Warren to the effect of "you guys better get your ass in gear because it looks like somebody might already be there." Warren was stunned. He went online and found the lecture. But the work of that scientist had not gone through peer review yet, and Warren realized eventually that it was not very convincing. "It turned out to be not too bad for us, but it was scary," Warren would recall later.

There were rumors of others getting close to using mRNA to make IPS cells. So Warren was desperate to use all of the resources of Harvard to bring their project to a close. This could send his career catapulting, but only if he were the first to publish. Rossi disagreed with Warren's strategy. He wanted to keep the work as confidential as possible. "Things were not looking good," Warren recalled. "I was aware that other people were possibly going to beat us to the punch, which was very scary."

Warren become so frustrated that he dashed off an angry email on November 11, 2009. "Dear Derrick, In light of the diminishing prospects that we can complete the RNA IPS project before the same goal is achieved by other groups and the continued deterioration in our professional relationship, I hereby submit my resignation."

It was an impulsive act. But Rossi still needed Warren to finish the project and publish the work. So he talked to Warren and convinced him to stay, promising to address some of his concerns. The crisis blew over within 24 hours.

Warren did not know it at the time, but he was just days away from finally succeeding. One day in November 2009, Warren was working in the lab when he realized he'd finally done it. He went to fetch Rossi. "I think we got it," he said. Rossi once again peered into the microscope. This time he saw a dish of IPS cells. "We did it," he told Warren.[16] From a purely scientific perspective, they had dramatically advanced the work of Yamanaka and found a way to make IPS cells for therapeutic purposes. It would, after all, land them in *Time* magazine soon.

Despite the celebration, the friction remained. Rossi wanted to publish the paper in the most prestigious journal possible, and for molecular biology that meant a journal called *Cell.* That journal has an impact factor of 32, meaning that the average article published was cited 32 times by other authors during the past two years. This put it in the same league with the most oft-cited journals. To make it in such a journal, the paper would have to be of great interest within the scientific community. It would also go through the most rigorous peer review process. Warren disagreed vehemently with the decision. He feared that the peer reviews at an elite journal would push back publication by months, if not a year. He was still afraid that others would publish before him and nullify his discovery. As Rossi recalls, "He thought, 'You're aiming too high. We're going to get very high bar questions.' And I was like, 'Well, I'm not worried by this. This is a really important study.'" Warren said he felt frustrated because Rossi was not listening to him. He contemplated resigning again. On March 30, 2010, Rossi submitted the article to *Cell.* He had an answer by April 24. *Cell* had rejected the paper, but it was now being considered by a sister journal, *Cell Stem Cell.* This was still a prestigious publication with an impact factor of 26.[17]

Even though *Cell* had rejected the paper, Rossi got back the comments from the journal reviewers. These are scientists chosen by the journal's editors to scrutinize the research for any weaknesses. The reviewers' identities are kept secret to prevent authors from trying to exert any pressure on them. Peer review is critical to giving scientific publications credibility. But the power reviewers hold over publication can be exasperating for the authors. "Reviewers always dig deep into their ass. They roll up their sleeves, and they reach as far into their asshole as they can possibly reach to pull out something that makes them sound clever, and it makes them think that they thought of something that you didn't think about. I can tell you, when we submit a paper after we've been working on it for several years, we've already thought about it," Rossi said. One of the three anonymous reviewers suggested that Rossi and Warren redo the most critical aspect of their research. The reviewer wanted to see if they could replicate the results they got in human cells if they redid the experiment using mouse cells. Warren was incensed. Trying to do the experiment again in mice, which he had already tried, would be a disaster.

The reviewer's comment seemed ridiculous to Rossi too. Why would anyone go back to mouse cells after proving you can revert human cells back to their pristine embryonic state? The whole point was to use IPS cells in humans, not mice. So Rossi called the editor-in-chief, Deborah Sweet, and told her that pleasing this one reviewer would waste 18 months and cost perhaps $150,000. "When it's done, is it really going to help get these cells to patients? No, because it's mouse experiments," Rossi said.

Two weeks later, Rossi resubmitted the paper with detailed responses to the reviewers. Rossi argued again against redoing the experiment. The journal weighed the revised draft but came back on May 24 with more questions. One reviewer had dug in his heels, Rossi told Warren in an email. Warren had had enough. He quit the same day.

"He was so intense and so paranoid that he thought he was going to get scooped," Rossi said. "He went off the deep end, as he is like to do, and basically quit. And at that point, he had already tried to quit a couple of times previous on the project, and I had successfully talked [him] back from the edge. But this time,

I thought, 'Well, you know what, we can do this without Luigi. It'd be great if he were part of the team, but if he really just can't handle this, then I'll let him quit.'"

As the back-and-forth over edits of the article went on, Rossi was preoccupied with starting a new company to cash in on their discovery. Although their work had discovered a better way to make IPS cells, Rossi insists that he always thought the most obvious use of modified mRNA was to treat genetic diseases. If someone inherited a faulty gene from their parents and was not producing a necessary protein, you could fix that—at least temporarily—by injecting them with mRNA that instructs cells to produce the correct protein. You would simply be supplying the genetic code that a person's body lacked. For years, drug companies had been making proteins in vats, purifying them, and injecting them into people. But the problem was that those proteins cannot get inside human cells, where they are often needed. Messenger RNA opened the possibility of treating thousands of diseases that are beyond the reach of protein therapy.

Rossi started plotting to meet with investors. In fact, the rumors of his discovery were spreading even before they were published in a journal. He would get calls from pharmaceutical companies who wanted to license the technology to make IPS cells. But Rossi would think to himself, "Don't these guys get it that we now have a technology that can produce any protein?" He held out for his own company. He had a more ambitious goal.

Rossi's first presentation of his lab's findings was a courtesy meeting with his colleagues. On April 14, 2010, he gave a talk at an interdepartmental meeting at the Immune Disease Institute, part of Boston Children's Hospital. It was obvious that Rossi and Warren's work brought Yamanaka's work closer to real-world applications. It created a buzz around the institute. One person who missed the meeting was Timothy Springer, a superstar faculty member who had founded the biotech company LeukoSite. That company, later acquired by Millennium Pharmaceuticals, would eventually have a blockbuster drug for treating colitis and Crohn's disease, a drug called Entyvio. The sale left Springer with about $100 million, making him one of those rare college professors who became extremely wealthy from his work. Springer had business savvy, having invested in another biotech. Technology transfer officer Ryan Dietz, who helps to sell discoveries by faculty, called Springer and asked if he'd be interested in talking to Rossi. Springer knew Rossi well and did not hesitate to agree to a meeting. Rossi and Dietz went to Springer's office on the medical school campus at 5 p.m. on April 27. They all sat at a round table in Springer's roomy office. Rossi had put together a PowerPoint slide show that was mostly about IPS cells. But it also listed using mRNA as a drug as one of a company's possible revenue sources. Springer considered himself a renaissance man of science with a wide range of knowledge. But modifying mRNA was an entirely new concept to him. Springer realized immediately that with stable mRNA, the possibilities were endless. "This is amazing. I want to invest," Springer said.

However, just three days later, Rossi and Dietz met on their own with a venture capital company, Third Rock Ventures. Once again, Rossi presented his slide show. The title page was daunting: "Reprogramming to pluripotency and directed differentiation using synthetic mRNA." The font was deep blue; the background plain white. The slide show was heavy on graphics, including slides of the cells glowing green. But the presentation was also heavily focused on Warren's IPS cell experiment. It was, after all, a breakthrough. And it would ultimately land Rossi on lists of notable people for the year in *Time* magazine. Rossi is skilled at explaining science in plain English. But he waited until the 25th slide of a 27-slide presentation to talk about the most critical issue in starting a new company: How to make money. Rossi listed three sources of revenue. The least lucrative was to sell kits for making mRNA to research labs. The next was to make customized IPS cells to test drugs on diseased human cells in a petri dish as opposed to using mice, which can be a poor proxy for humans in drug trials. Rossi thought this had serious potential. But the final source of revenue was therapeutics—using mRNA to make drugs inside the body. Rossi presented this as the most lucrative source, estimating revenues of $15 to $20 billion through 2025. Still, Third Rock wasn't convinced. It passed.

Rossi started realizing that he might have better success teaming up with people who knew how the world of venture capital worked. "Derrick felt he had a very hard time trying to get a company started," Springer recalled. "As an assistant professor at the time, he wasn't well known. But I knew he was a very smart guy and very capable."

Springer had a plan. He suggested they go talked to Bob Langer, a professor at MIT who had founded dozens of biotech companies. He's a billionaire today and may have been one when Rossi approached him. In any case, Langer was a legend. Having him involved would give Rossi's startup enormous credibility as well as business savvy. Langer was a chemical engineer who decided to skip the fossil fuel industry, where most people of his ilk went, and go into medicine. His first two projects as a young engineer would get him published in two of the most coveted journals, *Nature* and *Science*. One was to see if you could arrest the growth of cancer by cutting off the growth of its blood vessels. The other was to use plastic to see if you could slowly release drugs. Decades later, the blockbuster cancer drug Avastin would be approved by the FDA, although Langer would get nothing for it. Langer's reputation, however, would be cemented by his expertise in drug delivery. Given the challenges of delivery mRNA to cells, he could even offer real scientific insight.

Langer recalls that most of the May 25 meeting was spent talking about Rossi's IPS cell research. Langer thought that research had potential, but he was far more interested in using mRNA to turn humans into drug-making machines. Making proteins in a factory is expensive and time consuming. Using mRNA would dramatically speed up the ability to make drugs and allow you to treat diseases that were otherwise untreatable with drugs. That's because mRNA gets inside cells. Factory-made proteins do not. Many diseases are caused by a defective protein inside the cell, called intracellular proteins. Rossi had an approach that opened up so many new possibilities.

Langer later recalled, "I really thought this is a platform if we use it to make drugs that would be revolutionary. I thought it would be the best biotech company in history."

At the end of the two-hour meeting, Langer told Rossi, "This is amazing. What can I do to help?"

As Rossi left the meeting and crossed over the Charles River back to his laboratory, he was wondering what to do with Langer. He decided to bring him on as a cofounder, a bit unusual since Langer had not contributed any sweat equity, at least as far as the science went. But he knew that more than any scientist in America, Langer understood how to start companies. He also knew that Langer's expertise was in drug delivery, which could prove critical for mRNA. A few days later he asked Langer if he wanted to be a cofounder, and Langer said again, "Whatever I can do to help." With Springer and Langer at his side, Rossi had the clout that Drew Weissman and Katalin Karikó lacked.

Rossi also decided to bring in Kenneth Chien, another professor at Harvard, as a cofounder. Chien was a medical doctor and was working with Rossi on an experiment involving regrowing heart tissue in mice using mRNA. Rossi wanted as much firepower as he could muster, so having a doctor for a biotech company along with Langer, a biotech superstar, made a lot of sense. Springer would become an investor only.

In the meantime, Springer had also set up an afternoon meeting on May 28 with Doug Cole at Flagship Ventures. Springer knew Cole from the board of directors of another biotech in which he was an investor. That went well enough to schedule the next meeting with Noubar Afeyan.

Afeyan was by training a biochemical engineer, earning a PhD from MIT in 1987. Born in Lebanon to Armenian parents, his family escaped a war to emigrate to Canada. Afeyan was more interested in starting companies rather than working in a laboratory.[18] It all started when he went to a boring National Science Foundation conference while in graduate school. At lunch one day he sat next to an electrical engineer he did not recognize. But he still struck up a conversation. The engineer explained how he used his training to start a new company many years ago. That engineer happened to be David Packard, cofounder of Hewlett-Packard.[19]

Afeyan decided he would try to follow in Packard's footsteps but in his own field of biochemistry. He searched for classes at MIT that would help him use his scientific training to evaluate the potential of startup companies. His philosophy, developed over time, was to find companies that, instead of making incremental change, were transformational. At about the same time, he was considering whether to invest in a microbiome startup involved in researching fecal transplants to cure a severe bacterial infection.

Afeyan has a big personality and a big ego. Ultimately it would be his decision as to whether Flagship would invest in Rossi's idea. It wasn't an immediate decision. It would take months.

Luigi Warren was gone before *Cell Stem Cell* decided to accept the paper. By this point, relations between Rossi and Warren were at their worst. The two barely communicated, though Warren would still go out for drinks with others in the lab. *Cell Stem Cell* sent Rossi a congratulatory email on August 11, saying it had accepted the paper for publication. Rossi had managed to persuade the reviewers to greenlight the paper. Once a paper gets accepted, it goes to a status known as "in press." That means it has cleared all the hurdles except for final proofreading.

But in early September, something cataclysmic happened. The *Cell Stem Cell* editor, Deborah Sweet, privately raised questions about the integrity of the research. A whistleblower had contacted Sweet and said they had tried to replicate Warren's work but were unsuccessful. She called Rossi, who remembers "part of the rationale from this editor was, 'Well, we've heard that the lead author was sort of crazy and he doesn't work in the lab anymore.'" Sweet may have heard about his anthrax blog, which remained posted on the internet. Rossi was flabbergasted.

"This is outrageous," he told Sweet. This is not how scientific articles are supposed to be treated. The paper had already been accepted. It had already satisfied the reviewers' scrutiny. At this point, the way to challenge its findings was for others, after publication, to try to replicate the findings. That's how science works. Who could have possible challenged the work anyway? How would anyone know how to replicate Warren's method? Whoever it was must have somehow had the manuscript of the paper. That made Rossi suspect one of the peer reviewers.

Sweet had presented no tangible evidence of fraud. Rossi was confident that the data presented backed up the findings. Retractions of articles were rising in those days, but on average less than four papers out of 10,000 were retracted, according to watchdog publication *Retraction Watch*.[20] Rossi defended the work's integrity. "I never doubted Luigi's technical rigor and scientific ability or his honesty," he said. "As a scientist, he's really exceptional. Not good. He's exceptional." But Sweet insisted on one highly unusual condition to get the paper back in press: Rossi would need to get someone else to independently replicate Warren's findings.

Rossi sent an email to 12 of the collaborators on the paper. He omitted Warren, who at this point was consulting for the company Stemgent not far away. "As you know our paper was accepted for publication in *Cell Stem Cell* and was scheduled to appear online this week. However, due to a series of very strange and very extraordinary events (outside the domain of the science in the paper, and outside of the peer review process), publication of our story has been held up, and the editor has requested that we conduct additional experiments. As I said, the circumstances surrounding this decision are highly unusual. However, as many of you would agree, our study may be very important as it has the potential to transform multiple fields (including, and beyond, regenerative medicine). Considering the potential significance of this study, I think it in the best interests of the study to get these experiments done and then go to press."

Rossi had no choice but to let Warren know as well. As Warren recalls, "He said there's been an unwholesome development. A whistleblower had said this can't be replicated. And they are now unaccepting the paper until it's been replicated." Warren couldn't believe it. Such a thing almost never happens. Who would be in a position to try to replicate his work? Someone was questioning his honesty. This could put his career in jeopardy. And he had no idea who would challenge his integrity. What right did anyone have to accuse him of fraud without telling him who it was and upon what it was based?

Rossi would seek advice from Doug Melton, a mentor at the Harvard Stem Cell Institute. Both Melton and Rossi understood that the paper could usher in a whole new way to make therapeutics using mRNA. "They understood the huge power behind the platform," said Chad Cowan. "You can now deliver RNA and have it become a very powerful mechanism for doing stuff with." If the article became tainted somehow, that could potentially torpedo the groundbreaking science as well.

At a regular meeting of faculty at the Harvard Stem Cell Institute, one of the professors dropped a bombshell. Everyone knew that Rossi's paper had been held by the journal, a shocking development. Melton was there, along with Rossi's close friend Chad Cowan. Rossi wasn't affiliated with the institute and so didn't attend. Doug Melton asked the other professors if anyone knew anything about the issue. "I remember the point was we're all aware of this controversy, like what's going on?" Cowan said. "What the big deal?" That's when George Daley, a co-author on the paper, revealed that he had discovered a problem with the paper. According to Cowan, Daley said a summer intern in his laboratory, an undergraduate, had tried to replicate Warren's work and failed. There was a stunned silence in the room. "People's mouths were open," Cowan recalls. "They were like 'You're kidding! That's what gave you pause?'" Making induced pluripotent stem cells was advanced science. The fact that a student couldn't do it should come as no surprise. Why would a scientific journal halt publication for that?

Daley, a professor of biological chemistry, was a stem-cell superstar at Harvard. He published the first major paper following up on Yamanaka's discovery, which ironically ran in a more prestigious journal, *Nature*. He was an astute politician and years later, in 2017, became the dean of Harvard Medical School. Daley was on the scientific advisory boards of and held shares in three biotech companies, including the stem-cell company iPierian,[21] which was in the midst of raising $22 million in 2010.[22] Rossi, who was a junior faculty member, said he put Daley's name on the paper because of one suggestion he made. Daley was known to the stem-cell community, while Rossi wasn't. In a 2020 interview, Daley said that his role in the paper was very limited. Two other scientists working under him at the human stem cell core laboratory, Philip Manos and Thorsten Schlaeger, did some research and were coauthors on the paper. A postdoc in his laboratory spent time with Warren watching the process, Daley recalls. He said at the time he saw Warren's work as important for the producing of unmodified stem cells that were pristine and safer than Yamanaka had produced. He added, "I would be disingenuous to say I saw the huge opportunity that we're now seeing play out in potentially a world-saving vaccine."

Daley declined to comment on the whistleblower controversy during an interview in 2021. But a source familiar with his account said that Daley first heard about the whistleblower when the journal editor called him and told him. Although he asked around, Daley claims he never knew who the whistleblower was. Cowan laughed at that notion. He said Daley provided details of the young woman in his laboratory who couldn't reproduce the results.

Rossi didn't hear what Daley had said at the meeting. But in a conversation between the two of them, Daley threatened to take his name off the paper. Much worse, he also threatened to strip Philip Manos's data from the paper unless the experiment was replicated by the human stem cell core laboratory. This infuriated Rossi. Manos didn't work in Daley's laboratory. The data had been analyzed by the core laboratory. Rossi had paid for that work as an independent laboratory. Even though Daley technically oversaw that laboratory, Rossi said Daley shouldn't have had the power to muzzle its data, which was substantial and would have gutted the paper.

Rossi remembers holding another meeting that Warren attended. At it, Rossi discussed what they needed to do next. Warren was still in the dark about George Daley's role. There was a new researcher in Rossi's lab, Pankaj Mandal. Rossi went up to him and said, "'Pankaj, can you take on this job of generating IPS cells with Luigi's technology?' He had just joined the lab, and of course he's not going to say no." But Warren became very frustrated during the discussion.

Warren also had no idea that Daley might be behind the hold on his paper. He came up with a series of tactics to try to put pressure on Sweet to publish the article. The first was to ask Daley to talk to her. Given his reputation and clout, Warren thought that Daley could convince Sweet to dismiss the views of an anonymous whistleblower. Why wasn't he defending a paper on which he had collaborated? Rossi, who knew that Daley was threatening to withhold data, was opposed to the idea. Warren was annoyed. Why wasn't Rossi doing more to defend him? So, Warren decided on his own to shoot an email to Daley. "We already jumped through all the hoops," Warren said. "So what the fuck?" Warren's email was fiery. He wrote that he was baselessly being accused of fraud and Daley should insist that the journal publish the paper. Daley never responded.

The more Warren thought about what was happening, the more he thought "it was just like being assaulted." He could not understand why Rossi was not doing more. He even started to question whether Rossi might be the whistleblower. That made no sense, but Warren was frustrated. He decided to go "nuclear," as he called it. He sent a threatening email to Rossi as his last hope of getting him to pull the submission from *Cell Stem Cell* and resubmit to another journal. But his tactic was reckless.

"If Debbie Sweet does not reverse her position and/or identify her phantom whistleblower so that that person could be held accountable for their actions by the close of business today, I will refuse to sign off on publication in *Cell Stem Cell*," the email said. "If you decide to keep the manuscript at *Cell Stem Cell* without my permission, and therefore presumably without my name, I will be forced to pre-empt the publication with immediate self-publication of a monograph describing my work."

Warren had even prepared the stripped-down monograph with his name as the sole author. It was dated September 15. He offered other collaborators a chance to put their names on his paper. "If anybody involved in the validation essays wishes to be credited in the paper or offer their own data for inclusion, I will be happy to oblige," his email said. "You'll know where to look for it. Just Google my name."

It's hard to judge whether Warren was bluffing. Posting the monograph on the internet would have been a career-ending move. It would have undercut all the work he had done in the previous two years. It would have jeopardized publication of the work in any journal. Rossi cannot remember how he responded, but neither he nor Sweet met Warren's demands, and the monograph was never posted.

However, Warren was not done trying to apply pressure. Feeling powerless, Warren one day walked a few blocks from his new laboratory at MIT to the office of the editor of *Cell Stem Cell* in Kendall Square. He was in a huff. He had no appointment. He almost certainly could not have gotten one. He took the elevator up and barged in unannounced. He asked the receptionist to speak to the editor. "I knew it was kind of outrageous, but obviously I was very frustrated," Warren would recall later. Warren said it might have been apparent to the receptionist that he was annoyed, but he insists that he didn't make any threats. The receptionist told Warren she was not in the office. In any case, she never appeared. Warren vaguely recalled saying "fuck it" and walking out. He was only there a few minutes. Sweet notified Rossi, who had a flashback to his conversation with Warren about getting a gun. Warren says he did not bring the gun with him to Boston and chafes at the notion that anyone would think that he would have been carrying one. Sweet told Rossi that the office was so rattled that they closed for the rest of the day.

Given what had happened, Rossi notified everyone involved in the paper that Warren was unstable and to be careful. George Daley left a picture of Luigi Warren with security in his building and alerted local police, who put a cruiser in his driveway for two weeks. In fact, when Warren went to see a coauthor on the paper, Thorsten Schlaeger, on campus, a security guard confronted Warren and told him he wasn't allowed in the building, the same building where Daley had his office. The guard escorted Warren out.

By this point, other authors on the paper were raising questions about whether Warren's name should be left on the paper at all. One scientist who had worked closely with Warren, Philip Manos, argued that *he* should be made the first author. That position is highly coveted and can be a real career booster. It tends to signify the person who did the most work. Manos worked in the core laboratory that ultimately reported to Daley. Was Daley trying to elevate his status on the paper by having Manos get the credit? No one knows, but Rossi doubts it. The connection between Manos and Daley was not that strong. He didn't work directly in Daley's laboratory.

However, Rossi did elevate Manos's position on the paper. At some point before the controversy, Rossi stopped Warren in a stairwell to ask if he would mind if there were a footnote in the paper saying the Manos and Warren had "contributed equally to this work." Warren, slightly annoyed, said he did not mind. But that was a far cry from a request to pull Warren's name from the paper entirely. Rossi said he told Manos, "No way. I don't care how much trouble Luigi has been over the past year. He's going to get the credit for this because this is really his baby."

In the end, the new hire, Mandel, carefully went over Warren's process step by step. And it took only two weeks. The core laboratory confirmed the work. "He succeeded, and lo and behold, Luigi's science was real," Rossi said sarcastically. Rossi emailed Sweet to tell her of the results. "Here's the data generated by this kid in my lab—brand new fellow in the lab—totally independent of Luigi, that shows that you can generate modified RNAs that could be put into these human skin cells to turn them into IPS cells." He included the data and said he expected the paper to go back into press immediately.

Sweet sent an email back, perhaps by accident, that left Rossi speechless. It included a thread lower in the email with Sweet asking Daley if he was satisfied with the replicated data. For the first time, Rossi realized that Daley was the person who stopped the publication. The rules are that journal editors talk only to the corresponding author on a paper. It broke all the rules for Sweet to be talking to Daley. Rossi became irate. Daley talking to the editor behind his back was sabotage. In Rossi's mind, it was not just a betrayal; it was unethical. He does not believe that a respected scientific journal would halt publication of an article on the word of a college student in Daley's laboratory. Daley was the real culprit. He was the one who stopped the publication.

Rossi got on the phone with Daley and began screaming at him. "What fucking right do you have to do this?" he asked. "You're a senior Harvard guy, but I'm the corresponding author. What right do you have to do this, you son of a bitch?"

When Daley discovered in July 2021 that he might be identified as the whistleblower, he sent an email to Rossi. It said, "It pains me to think that you may have been harboring misconceptions about how the issue was raised. It was not by me. I'm wondering whether a phone call might be in order to clarify before they find their way into print."

Daley claimed that he himself did not know who the whistleblower was. But he said in the email it didn't matter. "With a whistleblower looming," the July 2021 email said, "I took the strong stand that we hold the paper to enable replication completely independently."

To Rossi, this was nothing more than a confession. "It's academically egregious," Rossi said of Daley's meddling. "He shouldn't have been talking to the editor at all," Rossi said, adding that Daley should have come to him with any concerns about the science. "There's a lot of unethical behavior in that."

A source commenting for this book confirmed the angry phone call. But Daley sees himself as making a heroic stand.

He feared what might happen if it were later discovered that the results were fake. In 2006, the stem cell community was rocked by two articles on cloning that the journal *Science* retracted as fraudulent. Daley could not afford to be associated with any article where there might be concerns about its integrity. Daley did threaten to withdraw his name as an author along with anyone from his laboratory. He also threatened to pull all his data, which would have made it difficult to publish. Daley wasn't surprised that a student in his lab might not be able to replicate an experiment. This was a new technique, and it could have been just a matter of a lack of technical skill. He dismissed Rossi's concerns that this could affect his efforts to start a biotech company. Daley asked Sweet to hold the paper until someone else could redo Warren's experiment. Daley recalls telling Rossi this before the paper was accepted for the second time.

The journal itself, *Cell Press Cell*, declined to respond to questions until asked by Daley himself to comment. "During the review process for this paper, a researcher contacted us with concerns about the reproducibility of the findings being described," the statement from the journal's press office said. "The researcher who raised this matter was not Dr. Daley or any of the other authors on the paper. As per our standard editorial policies, we contacted the authors, including Dr. Daley, to inform them about the concern, worked with them to investigate it, and then proceeded with publication."

Neither the anonymous source's account nor the journal's statement changes Rossi's mind at all. "It essentially proves that [Daley] was acting outside of the realm of what's academically acceptable," Rossi said. "He was essentially internally sabotaging our story." He maintains that it was inappropriate for the journal editor to be privately consulting Daley, even if his name was on the paper. He also denies ever talking to Daley about Moderna and said that by this point, the company had already been formed. The brouhaha over the paper, even if it became public, would have no impact.

The article was published on September 30, immediately after Rossi sent Sweet the new data. The crisis was averted. Despite Rossi's confidence that the conspiracy over the article would have had no impact on Moderna, who knows what would have happened? What if Warren had carried out his threat to publish it on the internet and no journal would publish it? As it was, Rossi would soon appear in *Time* magazine, and the discovery would be hailed not just within the scientific community but among mainstream news media. It is now considered a major contribution to IPS cell science. And it is the research that led to the founding of Moderna.

CHAPTER 6

Who Founded Moderna?

I wrote an in-depth article about the creation of the Moderna vaccine for a national newspaper in 2021. I had hoped to give readers behind-the-scenes access to the company's efforts as it was testing the vaccine. Initially the Moderna public relations officer showed interest. But then months went by with the same response. I was asked time and time again to explain exactly what my story would be. I suspected that this was a stall tactic. There was a tentative plan for me to talk to two individuals who were with the company from its inception: Robert Langer, a cofounder and a highly respected serial entrepreneur, and Noubar Afeyan, the chairman of Moderna. But even those interviews kept getting delayed. So I finally dashed off an email to Langer. In my mind, my message could not have been more benign: "I've already done some interviews about the scientific discoveries in the Rossi lab, and I know you and Timothy Springer were the first two people Derrick went to when he wanted to start a biotech [company]. So you have more insight than anyone on the potential of modified RNA both shortly after the initial discoveries and today. Would you have time in the next week or two to chat about these subjects?"

Somehow, this email touched a nerve at Moderna. The new chief corporate affairs officer, Ray Jordan, warned me on a video call that I had jeopardized any possible interviews at Moderna. This was not because I had reached out to Langer independently. He was, after all, only a director on the company's board. He has done countless media interviews and talks about both his life and the companies he has helped to found. The problem was that I had mentioned Rossi's name in the email. I was perplexed. I reminded Jordan that Rossi originally came up with the concept for Moderna and he was one of the founders. But Jordan shook his head and shot me a look like I had no idea what I was saying. He insisted that Moderna was not Rossi's idea.

This is not a trivial point. From the start, Moderna has constructed its own story of the company's origins, and in that version, Rossi plays a minuscule role, akin to a footnote. A search for Rossi's name on Moderna's website draws a blank. In their revisionist history, Moderna not only tosses Rossi's contributions to the side, but they also don't acknowledge the work of Karikó and Weissman. In Moderna's telling, Noubar Afeyan and CEO Stéphane Bancel crafted the concept for the company and solved the problems associated with mRNA. In one account, these obstacles were overcome during Moderna's first two years of "stealth mode"—a common dark period for startups. In a TED Talk in December 2013, Bancel gave his narrative of how the company began.

> So a few of us couple years ago, sitting in Cambridge, Massachusetts, thought about the following crazy idea: What if mRNA could be a drug? And the reason people have not developed mRNA drugs in the past, because from what I explained to you it's pretty obvious, mRNA drugs, why not? It's because of two things: mRNA creates

> an immune response. Why? Because a virus is made of mRNA like flu. So if we inject mRNA in a patient what happens? Your body thinks you just got the flu and it's not very good for a drug . . . The other reason is that mRNA was thought to be very unstable—minutes in vivo [inside the body]. I don't think a drug that you have to inject yourself every couple of minutes would be a good drug either. So those two reasons are really the foundations of why we never made mRNA drugs before. But we asked ourselves: What if? And we worked tremendously hard for two years, and I'm happy to report that we found a way. We have a way to work through those two barriers.[1]

The most dubious claim is that Moderna's own scientists solved the riddle of mRNA. Actually, Katalin Karikó and Drew Weissman made the groundbreaking discoveries. Weissman recounts with a sigh how he has seen Bancel say the same thing during presentations for scientists, even when Weissman was sitting in the audience. "Bancel has been saying that at scientific meetings for years, where he credits Moderna as inventing nucleoside modified mRNA. And all the researchers in the audience point out to him that that's not true . . . I think it's purely a marketing and an ego thing. So if they can market that they invented it, people will give them more money."

When I first shared the TED Talk with Rossi, he was incredulous, saying that the notion of scientists sitting in Cambridge coming up with the idea out of the blue is so ridiculous. Said Rossi, "Their whole modus is to rewrite the truth."

Afeyan is even more dismissive of the work of Rossi, Warren, Weissman, and Karikó. In a 2020 interview, Afeyan explained that as a venture capitalist, he hears pitches all the time and he's always on the prowl for companies that can change medical science. According to Afeyan, Rossi pitched a stem-cell company, an idea that Afeyan said he had heard and rejected before. Afeyan gives himself and Bob Langer credit for realizing during Rossi's pitch that there was another more promising path for mRNA. That meeting, according to the calendars of Springer and Rossi, happened on June 10, 2010.

> As I listened to this, given my experience and point of view . . . the field of stem cell biology and stem cells themselves as a therapeutic, I thought was not itself something that was going to be—at least at that time—a strong therapeutic approach. We had previously talked to many, many other people who are doing similar things with stem cells and we had over and over concluded that we did not want to start a company in the area. But in that meeting, started the concept, which came from Bob and my discussion about whether—quite separately from all that—one could apply RNA as a new drug modality in patients to make their own product and their own drug. And that notion, at least as of then, and the research we did for the three months after, had very little precedent. But what ensued was a several month process—where we do what we would typically do—is we went in house. I got a team of scientists within Flagship and you just asked the question. What if we could get patients to make their own drugs? And that kind of led us down a path of deciding that we should, in fact, form a company.

In short, Afeyan says the idea for Moderna was his. Timothy Springer, a Harvard professor and a Moderna investor who was at the meeting, noted that everyone in the room was smart and it was obvious from Rossi's work that mRNA could be used as a drug. "I think there may be some competition for who had the idea," Springer said.

One scientist who was a pioneer in mRNA research said the idea of using mRNA as a drug itself was so old that it probably cannot be traced to any one person. But it most certainly did not start with Moderna. Weissman and Karikó, for instance, intended to use mRNA as a drug too. And their discoveries made it possible.

Rossi flatly denies Afeyan's assertion that he proposed creating a stem-cell company only. There is ample documentation to prove that Rossi had using mRNA as a drug in mind before he ever met Afeyan. In the patent Rossi and Warren provisionally filed on April 16, 2010, they included the possibility of using mRNA as a treatment for genetic defects. "The modified, synthetic RNAs described herein can be used for the purposes of gene therapy," they said in the application.[2]

Beyond that, he made a case for using mRNA as a drug in the peer-reviewed article that was submitted to a scientific journal on March 30, 2010.[3]

In a YouTube video Boston Children's Hospital posted when Rossi and Warren's paper was published in September 2010, Rossi says in plain English what he said in the article: "In terms of therapeutics, any genetic disease that involves a mutation of a gene that doesn't make a certain protein we can now approach that with this technology to reintroduce that protein into those cells and re-establish proper function to those cells. So we think that this is going to be really important for many therapeutic avenues."[4] That was the core idea behind Moderna.

But the smoking gun, if one is even needed, is the PowerPoint presentation Rossi gave when he pitched the idea of Moderna. The file is date-stamped April 23, 2010, seven weeks before Afeyan met with Rossi. Most of the slides explained the research he and Warren had done, but the key slide was the one that broke down how to cash in on the research. Rossi estimated the company could make $15 billion to $20 billion over 15 years through the sale of therapeutics. As the slide shows, this included both stem-cell therapies as well as using mRNA as a drug. Moderna never pursued two other revenue ideas Rossi laid out, including one that involved using stem cells to help diagnose and research genetic disorders. Rossi said he never suggested the idea of using stem cells in transplants. He knew that was too far in the future.

Perhaps it is no surprise that anyone around when a company is started tries to take credit for the idea. But Afeyan does not stop there. He goes further to say that the modified mRNA science Rossi brought him was so deeply flawed that it would never have worked. Afeyan contends that Moderna ultimately had to fix Weissman and Karikó's science before Moderna could use mRNA in humans.

"In fact, at the time of Derrick's work and Weissman's, they had managed to reduce the innate immune response by 20 to 30%," Afeyan said. "And if we had gone in with those chemistries from 2010, which is not what we do today, and just shot it into people, you would have had a massive immune response."

This statement seems to reveal a colossal misunderstanding of the science at the core of Moderna, the scientists involved believe.

"That is nonsense," Weissman said. "Derrick's work would not have happened with a 20% reduction." Weissman said that after purifying the RNA and modifying it, he and Karikó eliminated any immune reaction. The data presented in Luigi Warren's journal article also undercuts Afeyan's claim. As a chart included in the article shows, Warren was able to either eliminate or dramatically reduce various immune responses in the cells. Rossi called Afeyan's statement absurd. "Noubar is not a scientist, by the way," Rossi said. "He's a professional self-enricher. And I would say liar. He's pretty good at that too." Warren did not mince words about Afeyan either. "He's full of shit. That's complete bullshit."

The experiments done at Penn and Harvard relied on laboratory-quality RNA. To make it safe for use in humans, more steps were needed for processes like purifying the RNA. This is typical when developing any medicine, but that had nothing to do with the science behind modifying RNA. Weissman pointed out that Pfizer's partner, BioNTech, used the formulations he and Karikó developed and made a vaccine that was effective at one-third the dose of Moderna's.

Afeyan seems unwilling to acknowledge that Moderna would have never existed without Weissman and Karikó's discoveries or without Rossi bringing that science to him. Whether they are willing to admit it or not, Afeyan and Bancel would be billions of dollars poorer today without those scientists.

So what's the true story of Moderna's founding? Afeyan did in fact assign two in-house scientists to evaluate the potential of mRNA. They were both 12-week fellows at Flagship, both just out of graduate school. Jason Schrum had just finished his graduate work in biological chemistry at Harvard University, where his advisor had been Nobel laureate Jack W. Szostak. Kenechi Ejebe had just completed medical school at George Washington University, an urban campus not far from the White House. Schrum and Ejebe worked for three months to produce a report on mRNA's potential. They looked at the available science to see if it was sound. They talked to scientists, including Rossi, Weissman, and Karikó. They also considered possible uses, market potential, and competitors. There were in fact already mRNA biotech companies. The German company CureVac was founded in 2000 by biologist Ingmar Hoerr, based in part on his graduate work in college.[5] CureVac focused on using mRNA as a therapeutic and a vaccine. "mRNA is like a memory stick," Hoerr once told a billionaire and potential investor. "You can just plug the memory stick into the body, it reads the information, makes any protein you want and the body cures itself." There were two other German companies only a couple of years old, Ethris and BioNTech. Ethris was focused on using mRNA to treat cystic fibrosis. BioNTech would of course go on to develop a COVID-19 vaccine in collaboration with Pfizer. Also in 2008, the UK company Shire Pharmaceuticals started work on using mRNA to treat cystic fibrosis.[6] Schrum and Ejebe started pulling together their notes to prepare a presentation.

Flagship also sent its vice president of intellectual property to meet with Weissman and Karikó in hopes of licensing their patent. But by the time Flagship's Gregory Sieczkiewicz, reached out to the pair, they no longer had any leverage to negotiate. Sieczkiewicz first insisted that Karikó sign a confidential disclosure agreement. The CDA she signed July 20, 2010, was to prevent her from talking about Moderna's efforts to acquire the license or to reveal a possible collaboration between Moderna and the Penn scientists. Although Karikó signed it, she told Sieczkiewicz that Penn had already sold the rights to Gary Dahl.

"Moderna would later claim that they discovered it. They never even heard about that we did modified RNA already," Karikó said. Sieczkiewicz had several meetings with Weissman and Karikó to seek their advice on mRNA. According to Weissman, Sieczkiewicz was still interested in whether to pursue IPS cells. But Weissman cautioned him against it. The real potential of modified mRNA was in vaccines and therapeutics.

"Flagship tried to bully Penn into licensing them the patent essentially for nothing. Penn refused to give it to them, and they got very upset," Weissman recalls. Penn had no choice. It had already licensed the technology to someone else. Flagship turned to Dahl to sublicense the patent from him. But Dahl said Flagship wasn't offering nearly enough, so he didn't bother to get into detailed discussions with Sieczkiewicz.

Sieczkiewicz became frustrated, Weissman recalled. At the last meeting, "He told us that they were going to form a company and that they were going to find a way to get around the patent." In fact, rather than fixing Weissman and Karikó's science, Moderna spent its first few months looking for another way to modify mRNA to get around the Penn patent.

In August, Schrum and Ejebe made their presentation to Flagship's board of directors. They rejected the idea of forming a company to produce IPS cells. Instead, they concluded that the most commercially viable use of mRNA was to treat rare genetic disorders, such as cystic fibrosis. That's also what the competition was doing. They did not include vaccines on the list because those didn't seem lucrative enough. Throughout the process, it always felt to Ejebe that the research was more of an academic exercise, perhaps because of his lack of experience working for a company. But despite his presentation, he never imagined that Moderna might ever become a successful company. Nevertheless, the report was well received. Flagship decided to invest in the new company that, for the moment at least, would be called LS18. Decoded, LS18 stood for the 18th life sciences company Flagship had created. Giving the company a number rather than a real name softened the sting if it didn't work out.

Still, Rossi had been working to come up with an actual name for the company. He almost settled on Harbinger, a person who signals the approach of someone. But as he thought about it, he realized the word originated from a person in medieval times who would ride to the village and warn them that an army was advancing. That seemed like a downer. He and Warren had called modified mRNA "mod RNA" for short. So Rossi had an idea. Why not add an *e* and make it ModeRNA?

When it came time to negotiate the terms of the startup, Rossi had the benefit of having biotech veterans Langer and Springer at the table with him. Springer wanted to be a 50% investor, but Afeyan wouldn't agree to that. He wanted full control as the company's primary investor. Flagship would put up two-thirds of the seed money and funding for the first few rounds. Springer would ultimately invest more than $5 million of his own money in the company, putting up $500,000 in late December 2010. He would add more money several times through 2012.

"We did some negotiation back and forth on what we thought was a good deal," Rossi said. "I really think it was because Bob was sitting at the table then we got really good terms." Rossi is skeptical that he could have done well in the negotiations on his own. Part of the deal was to give Flagship one-quarter of the founder shares, with the other three-quarters divided among Rossi, Langer, and Chien. Afeyan's argument was that he was going to have his own scientists work on the idea and Flagship would put in the sweat equity of improving the science. Rossi thought it was a bit unusual, but he later concluded that it meant Afeyan had a strong interest in protecting the founder shares. That ultimately proved valuable to Rossi.

Ejebe left Flagship for another assignment. But Schrum stayed and soon became Moderna's first and only employee for several months.

Just as Schrum began his work, Rossi and Warren's paper was published on September 30, 2010. As expected, the article garnered national press attention. There were stories in national and regional news outlets, including the *New York Times*, the *Boston Globe*, the *Chicago Tribune*, and *USA Today*. A piece on NPR began, "If doctors are ever going to use adult stem cells for medical treatments, they'll need to get a lot better at turning back the cellular clock and tricking regular old cells into becoming the multipurpose variety of their youth . . . Lead researcher Derrick Rossi told the *Washington Post* scientists now have a way to make 'patient-specific cells highly efficiently and safely and also taking those cells to clinically useful cell types.'"[7]

Even though LS18 was in stealth mode—operating in secret until it could distance itself from competitors—the company came out of the shadows to bask in the media attention. The press release even used Rossi's idea for a name. "ModeRNA Therapeutics, a new startup out of Cambridge, MA-based Flagship VentureLabs, announced it has developed a method for producing human induced pluripotent stem cells, which are embryonic-like stem cells that are formed by reprogramming adult stem cells," said one article. It was a bit ironic considering that Afeyan later said that he instantly rejected the notion of making stem-cells.[8]

The announcement went on to give Moderna's research team credit for developing modified RNA to reprogram cells back to embryonic stem cells.[9] Of course, it was really Rossi and Warren who did that. ModeRNA wasn't even incorporated yet and had only one employee, who was busy ordering supplies for his new rented laboratory.

The bigger splash came two months later when *Time* magazine listed Rossi as one of 50 "People Who Mattered." "Until now, the creation of these so-called induced pluripotent stem (iPS) cells, which are derived from adult cells instead of embryos, has required the use of potential cancer-causing viruses and genes to help coax the adult cells back into an embryonic state," the magazine said. "But Rossi's method bypasses that step, using messenger molecules—instead of the actual viruses or genes—to do the same work without the risks, such as cancer, that these elements pose."[10]

The new company once again took advantage of the publicity. "In its annual wrap-up of people and events that mattered, Time Magazine this year recognized the work behind ModeRNA Therapeutics," a press release said. "The magazine also cited Dr. Derrick Rossi, Asst. Professor of Pathology, Harvard Medical School, for his role in leading the team that developed the breakthrough process."[11] Again, *Time* never mentioned Moderna.

Working on his own, Jason Schrum spent much of his time in the first few months of Moderna's existence looking for a different way to modify mRNA. The company had no choice because it had been unable to license Weissman and Karikó's patent. Schrum did not have an actual office. Flagship subleased a lab space in Cambridge's Kendall Square, not far from MIT. The lab was a bit bleak, Schrum recalls with a laugh. It had no window. There was a single bench and a hood to vent fumes. He was isolated even from the company subleasing the space. The only breaks from his isolation were occasional visits from Sieczkiewicz.

Schrum's background was in nucleic acid chemistry, and he had even done research on modified nucleosides for another purpose. While Weissman and Karikó proved that modifying mRNA allows it to evade the body's immune reaction, there are many ways to modify mRNA. Finding a different way might produce better results and, critically, allow Moderna to get around Weissman and Karikó's patent.

Schrum didn't really consider his work to be groundbreaking. It was called a pilot study. He had a good idea from his training as to what might work best. And he was right. Weissman and Karikó had looked only at nucleosides naturally found in humans, which enabled Schrum to test a lot of synthetic or unnatural nucleosides. Schrum tried a list of nucleosides that he thought might do well, including some he had to design himself and custom order. When he ran the first pilot experiments, some of the modifications worked. But Schrum wasn't surprised. It was what he had expected. The one that worked the best, n1-methylpseudouridine, was so effective at making proteins that it blew out Schrum's ability to measure the results. Rather than jumping for joy, Schrum was annoyed because he was going to have to do it again and dilute the mixture. It was going to take at least another week. This was during the winter of 2011. Schrum found that n1-methylpseudouridine worked better than the modification Weissman and Karikó had done. It created less of an immune reaction while also allowing the mRNA to last days longer and to produce more proteins. That is the modification that was ultimately used in the mRNA COVID-19 vaccines. To this day, a more effective modification has not been found.

Schrum filed for a provisional patent even before Moderna had been incorporated. The company later used Schrum's employment letter to assign the patent to Moderna. Curiously, both Stéphane Bancel and Noubar Afeyan were also listed as inventors on the revised patent application, even though neither had spent time in the laboratory.

Moderna would later present Schrum's discovery as the core science behind the company's genesis. It would describe Weissman and Karikó's work as only inspirational. Schrum would be referred to as Flagship's "innovation team." It was in fact a team of one. Here's Moderna's official account of his beginnings in a Securities and Exchange Commission document:

> Moderna was founded in 2010 by Flagship Pioneering to develop and commercialize a new category of medicines to treat human diseases . . . Inspired by chemically-modified mRNA used in cell culture experiments, the [VentureLabs] innovation team, working with a team of scientists assembled to launch Moderna, identified chemical modifications of mRNA, engineered mRNA sequences for greater in vivo potency, and demonstrated our first instances of in vivo protein expression.[12]

Schrum's work had been impressive, but as Rossi would later put it, "The fact of the matter is that [Weissman and Karikó] and not [Moderna] demonstrated that nucleoside modification is the key to getting mRNA into cells to get proteins expressed. That is the key discovery. That is the Nobel Prize–winning discovery, and I do believe, and I surely hope that Karikó and Weissman will get the Nobel Prize because I think they deserve it for that discovery."

Schrum agrees that Moderna owes its existence to the discoveries of Weissman and Karikó. "Moderna wouldn't have been an idea if Luigi Warren hadn't used their technology on the Yamanaka factors," Schrum said. "It just wouldn't have been a thing."

Weissman and Karikó's original patent included broad enough language that Schrum's modification was included in it, according to Weissman. Moderna efforts to elude that patent ultimately failed. The company in June 2017 went back to Cellscript and sublicensed the intellectual rights to Weissman and Karikó's discovery. BioNTech, the other maker of a successful COVID-19 vaccine, would do the same.[13] Moderna paid upfront grant fees of $75 million and well as additional fees and royalties.[14] In the first quarter of 2021 alone, Moderna paid Cellscript $84 million for the license.[15] The patent that the University of Pennsylvania would not license to its own inventors and gave away for a pittance is now worth hundreds of millions of dollars. Penn splits some royalties with Weissman and Karikó, reportedly worth millions of dollars. But while several people became billionaires from Karikó and Weissman's breakthrough, they did not.

"The University of Pennsylvania, their tech transfer office is impossible to work with," recalls Schrum. "What they did, which was incredibly foolish, was to give an exclusive license of the use of these modified nucleotides to a research tools company."

CHAPTER 7

Enter Stéphane Bancel

If corporations are people, Moderna is Stéphane Bancel. He joined the company's board in March 2011 and officially became its chief executive officer seven months later. Bancel has made Moderna the successful, cash-rich company it is today, and Moderna has made Bancel a billionaire. As of September 2021, his shares in the company alone were worth $12 billion. And that was after he raised a few eyebrows by selling $154 million worth of his stock starting in early 2020. An insider selling stock, especially at such as critical time, can raise questions about the seller's confidence in the company's value. Bancel is a more of an extraordinary dealmaker and salesman than a scientist. People who know him say his real genius is raising money. Moderna became a unicorn, raising more than $1 billion, years before it had any product on the market. Arguably, if Bancel did not have this talent, the company might not have an mRNA vaccine today.

In public, Bancel comes across as articulate and charming. Behind the scenes, many find him to be profane, ruthless, and egotistical. "I'm the type of person who believes in what I believe and doesn't really care what other people believe," Bancel once said.[1] One of Moderna's cofounders, Kenneth Chien, told a writer for the French version of *Vanity Fair*: "You are going to hear a lot of things about Stéphane. Good: a visionary, a genius somewhere between Jeff Bezos and Elon Musk. And at least not so good: a bubbling, arrogant, unbearable boss."[2]

From its earliest days, outsiders' views of Moderna were wildly mixed. There were those who thought the company would one day revolutionize medicine. There were those who thought Moderna was all hype. For years it was hard to tell what was going on behind closed doors because the company was so secretive. It preferred touting financial accomplishments rather than its science. In a scathing investigation by *STAT* in 2016, the respected health publication said it "found that the company's caustic work environment has for years driven away top talent and that behind its obsession with secrecy, there are signs Moderna has run into roadblocks with its most ambitious projects . . . Interviews with more than 20 current and former employees and associates suggest Bancel has hampered progress at Moderna because of his ego, his need to assert control and his impatience with the setbacks that are an inevitable part of science. Moderna is worth more than any other private biotech in the US, and former employees said they felt that Bancel prized the company's ever-increasing valuation, now approaching $5 billion, over its science."[3]

When *STAT* asked Bancel about his reputation as a terrible boss, he said he had been seeking tips from Silicon Valley companies, such as Facebook, Google, and Netflix. But he made no apologies.

"You want to be the guy who's going to fail them? I don't," he told *STAT*. "So was it an intense place? It was. And do I feel sorry about it? No."[4]

Bancel grew up a child of privilege. He was born in 1972 and raised in Marseille, France, a port city on the Mediterranean. His mother was a doctor. His father was an engineer as well as a sailing junkie. Bancel told an interviewer once that the first picture of him as an infant outside the hospital was in a bassinet on a sailboat.[5]

Bancel grew up boating in the Mediterranean some weekends and other weekends driving two hours to the Alps to go skiing. He spent a lot of free time in Italy and Spain.

As a teenager, he liked to write computer code on his Apple IIc, which came out in 1984 and cost a pricey $3,200 in today's dollars. He admits to making Cs and Ds in biology in high school, perhaps as a way of rebelling against his mother's profession. It did not make her happy.[6] So he decided to go into engineering, his father's profession. He also loved chemistry. After college, he earned a master's degree in engineering from École Centrale Paris and the University of Minnesota, where he focused on biochemical engineering. He said working in a laboratory, he realized he hated research. He was too impatient to wait for results.[7]

Between graduate schools, Bancel did an internship in Japan. He was able to avoid compulsory military service while in school, but once he graduated, he had to find a job with a French company. So he got a list of all life science companies in Japan from the Chamber of Commerce and applied to at least 10 of them. He was able to land an internship at the French diagnostic company BioMérieux in Tokyo, where he worked for free.[8] He fell in love with the country. He even became fluent in the language. So after getting his second master's degree in Minnesota, he decided to go back to Japan. He found a job there again with BioMérieux. Despite having two graduate degrees, Bancel became a sales representative, selling medical equipment to quality-control laboratories for food and pharmaceutical companies. The equipment made sure dangerous bacteria did not end up in food and drugs. He eventually became the head of sales in Japan, and later he oversaw the industrial microbiology business for the Asia-Pacific region.[9] After four years at BioMérieux, Bancel decided to go back to school in 1998, this time to study business. He was accepted to the MBA program at Harvard Business School. After backpacking in Australia, he headed to Cambridge, Massachusetts.[10]

Bancel graduated with an MBA in May 2000, just after the dot-com bubble burst. Startup companies with huge losses but lots of hype—the so-called new economy that was divorced from quality products and healthy profits—were exposed as grossly overvalued. Still, many of Bancel's classmates wanted to go into the now deflated worlds of finance and technology. Bancel was more interested in pharmaceuticals. Recruiters from Pfizer, Medtronic, and Eli Lilly came to the Harvard campus, but Bancel recalls hardly anyone else was meeting with them. Lilly considered Bancel for a marketing job, but he insisted he wanted to go into manufacturing instead. He had heard for years that the biggest problem drug companies face is in manufacturing quality control. Many companies had run into trouble with the FDA after a plant inspection, trouble that could cost them dearly. Bancel later said, "It's crazy to spend so much time in the labs making those medicines if you cannot find a way to make them with the right quality to get them to patients."[11] So he worked in a drug plant in the United Kingdom for a while.

"I actually liked it because again I was playing a long game," Bancel told an interviewer in 2016. "I was just thinking 'Geez it's going to be so funny in fact 10 years from now when people realize that I'm playing chess, not checkers, and that I'm going to be having a great life and a great career.'"[12]

Suddenly, there was a problem at a plant in Indianapolis. During a routine inspection, the FDA found problems at a factory about to produce a new rapid-action formulation of Zyprexa, a drug used to treat schizophrenia, bipolar disorder, and depression.[13] Bancel was called in to help. The FDA had issued a warning letter for problems at a site that made injectable insulin. He spent two years helping the team fix the problems before deciding it was time to go into the corporate world of the giant pharmaceutical company. He eventually moved to Belgium to become a culture manager.[14]

One day, unexpectedly, Bancel got a call from Alain Mérieux, the chairman of BioMérieux. Their relationship by then was limited mostly to exchanging Christmas cards. Mérieux said he was visiting family in Brussels and wondered if Bancel had time to meet. He took Bancel to a cafeteria for lunch, where they talked about French politics and business. Just before dessert, Mérieux dropped the bombshell. "We are looking for a new CEO, and I've been advised to throw a very wide net. Of course, you're way too young."

Bancel was only 33 at the time. Even so, he landed the job. Mérieux credited it to Bancel's unique experience in manufacturing. Like Lilly, BioMérieux had received an FDA warning letter for a plant in the United States. Bancel researched the problem and concluded that it was fixable. BioMérieux made products that diagnose diseases, including bacterial and viral infections. It was a global company with four presidents. It had teams in 42 counties with 10 plants and 10 laboratories doing research and development. In 2006, the year Bancel started, it had $1.37 billion in sales.

Bancel admits he was intimidated shortly after accepting the job. "I was kind of very happy for one or two evenings, and then really scared for a few months. Because (I was thinking) 'Oh geez, what am I going to do with my life?'" There were so many aspects of the business for which he had no expertise, which led to sleepless nights.[15]

When Bancel arrived at BioMérieux in July 2006, he focused intensely on the company's market share. For many years, BioMérieux had 30% of the global market, about the same as its fiercest rival, New Jersey–based Becton, Dickinson and Co., better known as BD. At his first board meeting, he told the directors, "In five years, we're going to have 40% market share." He says no one believed him because market share had been flat for 15 years. "We're going to do it by pushing harder the current product, doing better launches, putting more money in R&D and buying companies." Despite skepticism within the company, Bancel told the same thing to Wall Street analysts, making his pledge public. "But then everybody in the company was terrified," he said in a later interview. "Someone will ask me, 'How are you going to do this?' 'I have no idea. But you guys are going to tell me.'" At every meeting in every department he would ask one question: How are we going to get to 40% market share? In retelling this story, Bancel says the company got to 42% market share within three years.[16]

Much of what Bancel did at BioMérieux was to make deals. He would fuel growth with new acquisitions to fill holes in the company's R&D. He told Reuters in 2008 that he preferred smaller companies with innovative products rather than buying bigger companies with a lot of duplication of services.[17] In his five years at BioMérieux, Bancel bought 10 companies, including a couple he regretted. That brought him into contact with venture capitalists who began to ask if he wanted to run a company.[18] During this time, he would have occasional discussions with Noubar Afeyan at Flagship. Afeyan would finally convince Bancel to become chairman of the board of one of Flagship's portfolio companies, BG Medicine Inc., in 2011.[19] But prior to that, Bancel would get calls from other venture capitalists. He estimated that he turned down perhaps a dozen feelers.

"I got to know all the best VCs in the US, including Flagship. Some of them started to propose to me to run some of their portfolio companies," Bancel told an interviewer. "And they were all one-drug companies. I'm not a gambler. I always politely say, 'Thank you very much but no.' Because I know that most drugs never get to market."[20]

Bancel was ready to leave BioMérieux, however. There were two big acquisitions that he tried to engineer "that the board at the 11th hour kind of freaked out and didn't do. And after the second one, I thought, 'Okay, I'm done.'"[21] He started interviewing at other big pharma companies. He got an offer to run Hospira, a Chicago-based maker of injectable drugs. It was twice the size of BioMérieux, and it offered a way out. He flew home on a Friday night, and his wife suggested they go out and celebrate. But Bancel was not in the mood. Something bothered him about the job. He told his wife, "Look, it's going to be the same thing all over again . . . I'm going to spend the first three months on a plane going around the world, talking to people and diagnosing the company. I'm going to have to figure out who do I keep on the team. I'm going to have to fire people, replace people. And then I'm going to have to agree with the board on a new strategy." Then what was he going to do? Spend his time again buying other companies? It did not feel like the right challenge. He turned it down.[22]

But then he got a call that piqued his interest.

"One day Noubar Afeyan called me and said, "Where are you?' I said, 'Actually, I'm in Boston. Why?' 'Well, come to Flagship tonight. 6 p.m.' So I arrived at his office and it's dark outside, February or whatever. And he showed me an experiment—one experiment with 10 mice. He showed me mRNA that had been injected in the muscle. Human EPO. And you can find human EPO in the blood. Meaning we turned the mouse into biotech factories . . . He showed me the data and he said, 'What do you think?'

"I'm like, 'This is not possible.' And he said, 'What do you mean?' [I said,] 'This is one of those things that happened every day of a week from academia, you know, a nice paper but you cannot replicate the data. So I'm like, 'Thank you very much for wasting my evening. I'm going to go have dinner with my wife and kids now.' And he's like, 'Stéphane, don't go now. Just for our friendship, you will have a lot of time to check the data. Assume for five minutes that this is for real . . . ' And we spent an hour going crazy."[23]

As they discussed it, Bancel said he realized all the advantages of mRNA, one of which is its simplicity. Messenger RNA is often referred to as the software of life. Much as everything digital can be reduced to a zero or a one, mRNA is made up of four letters in different sequences. Those four letters keep us alive. As he thought about it, Bancel concluded that if you could prove that one drug with mRNA worked, you could make almost any drug. What's inside the syringe wouldn't change. Only the sequence of those four letters would change. As Bancel would later explain, "You know if it was going to work for one, it was going to work for many."[24]

That wasn't the only advantage of mRNA. Drug companies use recombinant proteins in medicine to treat disease, but their drugs are limited to proteins outside our cells. Most proteins that our natural mRNA makes remain locked inside the cell. So that opened up a large catalog of diseases the startup company could treat that couldn't be treated with conventional drugs. "We have 22,000 proteins in our bodies, and two-thirds are undruggable, meaning you cannot make a drug for them," Bancel said. Messenger RNA would change that.[25]

That was not all. Drug companies devote enormous effort to filing patents on discoveries that could lead to new drugs. So in all likelihood, all useful recombinant proteins have already been patented. But mRNA is not a protein. It is what DNA uses to make proteins. That opened the door to filing a slew of new patents for RNA. "So all the patents from the old biotech industry most probably are useless, meaning you can go and do anything you want," Bancel said.[26]

Bancel left the meeting with Afeyan and walked across the Longfellow Bridge over the Charles River to get back home. "I was almost half drunk because my brain was just racing," he said. "I didn't sleep for a few nights just thinking about the possibility." He took a couple days off from BioMérieux to visit Derrick Rossi's laboratory at Boston Children's Hospital.

The experiment Afeyan showed Bancel that night had been done in Rossi's lab. After proving that you could induce cells in a petri dish to make proteins with mRNA, Rossi had decided to see what would happen if you tried it in a live animal. This is a common sequence in biological research. First you test something in vitro (outside of a living organism), and then move on to in vivo (inside a living organism). One of Rossi's graduate students, Wataru Ebina, started by getting the genetic sequence for a protein found in fireflies. Known as luciferase, this protein glows in the dark, which is what allows fireflies to set off those flashes of light in the dark. He coded the mRNA to make this protein. Then he anesthetized the mice and injected them with the mRNA. The mRNA interacts with another substance known as a substrate that had also been given to the mice. Ebina moved the mice to a pitch-black box with a light-sensitive camera and waited. Over time, Ebina could see the thighs glowing. This is pure scientific research. The only point was to prove that you could express proteins in animals using mRNA.

Getting a mouse thigh to light up is revealing but has no practical use. Rossi now wanted to see if he could produce a therapeutic protein in mice. This time, his graduate student encoded the mRNA to make a protein known as erythropoietin.

If Rossi were able to show that he could get mice to produce human EPO with mRNA, it would lend serious credence to the idea that you could use mRNA as a drug. The student was able to get the 10 mice he injected to make human EPO. Blood tests proved it. This is the experiment that Afeyan showed Bancel at Flagship, the one that Bancel said was impossible and could never be replicated. It's not clear why he would have thought this. Making proteins is the basic function of mRNA. Rossi never published the results of that experiment, but he did present it to the Moderna board of directors, just to show that mRNA could be used to make drugs in animals. "With every new modified RNA we made, we realized we could apply it to anything in biology," Rossi would recall later. "It all hearkened back to DNA makes RNA makes protein makes life. We now had mRNA in our hands, so we had the key to making life. The possibilities were endless." Afeyan was clearly impressed.

Bancel was excited about Rossi's work and became convinced the red blood cell experiment was not just a fluke. He also talked to Bob Langer, one of four cofounders, who told him that he thought mRNA could revolutionize medical science.[27]

Bancel continued to ponder what to do for a few days. He later said in an interview, "So if I say no, and this works, I'm going to hate myself the rest of my life." Bancel's wife, a photographer who knows little about science, finally said, "You have to do it . . . This could change the world." If that were case, she added, somebody was going to have to make sure the company succeeds. "So you have to go make it work." Bancel was convinced. They went out to celebrate in a bar over a bottle of wine.[28]

Bancel says if all he cared about was money, he could have taken safer jobs with big pharma companies. He says he joined a startup company with one employee because he saw the potential to change medicine forever. When Afeyan asked him to accept the premise that you could use mRNA to make drugs, Bancel started ticking off the ramifications. "I said, 'If this is true, you're going to be able to do hundreds, maybe thousands of drugs that are undoable using pharma small molecule or biotech large molecule proteins. You're going to be able to have a much higher probability of the drugs getting approval.'" He asked Afeyan, "By the way, how do you make mRNA?" Afeyan explained, "Oh, you're going to get a kick out of this. It's cell-free in water with enzyme." Bancel replied, "Shit, it's going to be super cheap."[29]

Bancel was named to the board of Moderna in March 2011, giving him time to settle affairs at BioMérieux. He would leave the French company in July to become executive chairman of Moderna and a senior partner at Flagship. He would take the role of CEO in October.[30]

Years later, in an interview with TED Talks, Bancel said he was already very wealthy when he went to Moderna.

INTERVIEWER: When you were done running (BioMérieux) you could have just bought an island and done nothing. Is that true . . . or not true?
BANCEL: Kind of true.
INTERVIEWER: So you didn't need to work another day in your life. You'd help a lot of people in the health space and you decided to jump in. Is that right?
BANCEL: Yeah.[31]

As a cofounder and a board member, Rossi played a role in hiring Bancel. His first impressions were favorable. Bancel seemed like a good candidate. Even years later, Rossi gave Bancel credit for having excellent business savvy. He was a genius at raising money. Rossi had a quarter of the original founder shares. Robert Langer had the same amount, along with additional shares for serving on the board of directors, worth $4.4 billion in September 2021.[32] Rossi will not confirm whether he is also a billionaire, but he acknowledges that Bancel helped make his shares worth hundreds of millions of dollars.

But Rossi was stunned when Bancel wanted Rossi to sign an agreement that all experiments he did in his laboratory would become the property of Moderna. He worked for Boston Children's Hospital, who paid for his research. Everything he did there belonged to the hospital. Everybody in science understands this. This is the rule of academic science. Even if he wanted to, Rossi couldn't give away his academic research, not even to his own company. However, Rossi remained diplomatic. He told Bancel that he could use his scientific knowledge to keep the company informed, but he couldn't give away his research for free. Moderna, in fact, licensed the patent Rossi and Warren were awarded for their stem-cell discovery. Looking back on it, Rossi still cannot believe Bancel even suggested it. "You're kind of stealing from Boston Children's Hospital, a hospital for children," he said. Rossi began to develop a strong dislike for both Bancel and Afeyan.

"They're really awful people. I mean truly, truly awful," Rossi said. "In the four years I was on the scientific advisory board of Moderna and there every week talking to their scientific team, I never once heard [Bancel] use the word 'patients.' It was all money, money money. That's the only thing he's interested in."

A mentor to Rossi said, "Derrick would come to my office for what I would almost call psychotherapy. He was so upset with the way Stéphane treated him. Let's just say that relationship became beyond dysfunctional. It was that Stéphane decided, in some sense correctly, that Derrick's expertise had nothing to do with the clinic or anything and so he didn't need him anymore." That's a theme repeated by others who worked at Moderna.

Even after Bancel agreed to take the job at Moderna, Schrum remained the only employee for a while. In early 2011, the company of one employee was moved to a new space in Kendall Square. This one included an office and large windows, an upgrade from the windowless lab with a bench. He began to hire other scientists.

Bancel's job at BioMérieux was to figure out how to grow the company. Now he was at a startup, but he had many of the same goals: rapidly growing and rapidly monetizing Moderna. He quickly developed a reputation as someone who did not care about the challenges his workers faced. Doing experiments with mRNA often involved coming back to the lab at odd hours—even in the middle of the night—to check on how long the mRNA lasts and how much protein it has produced. These were called time points, and someone had to keep track of them. As a result,

Schrum said he was working all the time. Yet one morning, when he came in around 8 or 9 a.m., Bancel was upset. He stopped Schrum and said workers at the company next door arrived at 7 a.m. Why was he coming in so late? Bancel once told a journalist that he himself only got three or four hours of sleep each night.[33] He had the same complaint with another newer employee, Suhaib Siddiqi. He lived 12 miles away in Burlington, which because of traffic can be a long commute. Once he arrived at Moderna at 8:15 in the morning and Bancel lectured him.

"When you have a family, work is not slavery. You have to drop your kids off at school. You can get stuck in traffic," Siddiqi said. "From that point on, I thought, okay, there's a problem with this person."

In public settings, Bancel could be quite charming. One employee remembers once when elevator doors opened, Bancel insisted that everyone wait for the scientists to exit first as a sign of respect. "Without you, we cannot do anything," Bancel said. But his demeanor could change behind closed doors. Bancel also became known for setting unrealistic expectations. He'd ask the scientists how long a project might take. According to Siddiqi, if the answer was four weeks, Bancel would say, "What the fuck are you talking about? I need papers on my table in one week."

Schrum recalls that as Bancel spent more time with the staff, he had to scale down his expectations. "He had the kind of large company sort of expectations and now . . . there were like five or six or seven of us and that was it. You don't have hundreds of people."

One of the first tasks assigned by Bancel was to file as many patents as possible. "Let me just give you a window into Stéphane's mind," one early employee said. "Once Stéphane learned that that there's over 20,000 functional proteins in the human body, we were basically charged with producing all of them." It was not about science. It was about business. Bancel wanted to prevent any competitor from patenting any mRNA coded for a protein before Moderna, even if no one knew yet what they would ultimately do with that protein. Officially, this plan was called Project 800. The number in the name says it all.

Bancel was impatient about doing the work needed to file the patents. To buy the supplies and make a protein could take up to six weeks. Even if the scientists already had the basic supplies, they still had to order the mRNA encoded for a protein, which took at least a week. Then it took at least a couple of days of make a protein and verify it for quality control. All told, it could take a week and a half to show that mRNA could produce a protein. But Bancel talked as though this work could be done in just a few days. "That's putting the scientist in a situation where they are forced to cut corners," Siddiqi said. "Not only to cut corners where it was possible but to cut corners where it was not possible just to save their neck and their job."

Scientists were annoyed that even though Bancel did none of the work, he would insist on getting credit for each patent. Said Siddiqi, "He's not a scientist although he's on every patent. He forces the company to put his name on the patent."

Employees left scathing anonymous reviews of Moderna on the job website Glassdoor in the company's first three or four years. In review after review, the employees would give the company one star out of five overall. The categories included work/life balance, work culture, and senior management. The comments were consistent. "The workloads are outrageous," said one from October 2012. "Bad science practices, blame and finger pointing abound." "This place is a disaster," said another from the same month. "The CEO and CSO [chief scientific officer] have no respect for the employees. It was made clear from the beginning that everyone is completely replaceable." Another from May 2013 said, "The management is extremely arrogant and believes everyone beneath them is replaceable. The CEO runs the company like a dictatorship and shows absolutely no respect for anyone but he demands everyone respect him." From August 2014: "Stressful, toxic atmosphere that brings out the worst in everyone. Goddamn awful management team whose plans change daily, who talk all the time about the Moderna 'family' in our employee meetings but then fire people on a whim." One review in September 2014 from one of the company's directors of chemistry said, "Honestly speaking, I had never experienced such an abusive, manipulative and arrogant CEO and the-then-CSO at any company in my entire life. The behavior from management was to blame everything on others."[34] After several years, there were finally some five-star reviews. But several former employees confided that they are skeptical that the glowing reviews on the site are truly independent.

Scientists started noticing that people would get fired not long before their stock options were supposed to vest. By not doling out those shares, Bancel would keep existing shares from getting diluted. That meant more money for him. "Stéphane used to obsessively look at the cap table—the [market] capitalization table—and try to come up with ways to get all of the equity back to himself. So he would actively look at how to not have to deplete or dilute equity that he thought should have been his own," one employee said. "People were fired like the day before the vesting date. Things like that that would happen regularly."

Siddiqui said one day he was supposed to meet with the chief scientific officer first thing in the morning. But the meeting was postponed and moved to another room. When he arrived, there was a woman who handled human resources with papers on the table. The chief scientific officer walked into the room, sat down, and said, "I don't think you are qualified as a chemist." Siddiqui did not hold back. He told his boss that he was not qualified to be a chief scientific officer. Just to put a punction mark on it, Siddiqui told the CSO that he would never hire him as a research assistant. Siddiqui was fired on the spot. He left the building immediately.

Not long after that, Jason Schrum left. Shrum had confided to Justin Quinn, the second scientist to work at Moderna, that he felt like Bancel might be trying to push him out. "They had already hired what he perceived to be his replacement and that turned out to be fairly true," Quinn said. Schrum was highly respected, and his departure stunned many of the scientists left, some of whom he had hired.

Quinn was next. He made a mistake one day that got him in trouble. When he produced RNA from a DNA template, he was supposed to use an enzyme to chew up the DNA to avoid any reaction. On paper, it looked like he didn't do it, but Quinn had an excuse. "I was working my butt off and wasn't taking very good notes," he said. He was asked about it and gave his explanation to the chief scientific officer. Later, the CSO asked him to bring his laptop to a meeting at 2 p.m. Quinn assumed he was going to explain once again what happened. When he walked in, Bancel's assistant, who also handled HR matters, was there. The scientific officer said, "It's not working out. We're letting you go." Quinn was crushed. "There was a bit of a misunderstanding and they just used that as an excuse to get rid of me," Quinn said later. "I feel it was not that big of a deal. You should not get rid of me even if I did skip the DNA step." A colleague said Quinn was close to his one-year anniversary, when his stock options would have vested. "They did not offer Justin any opportunity to remedy, just fired him immediately." Quinn was under the impression that he was Bancel's favorite employee. "He was always telling me stories about his youth that he wasn't telling anybody else. And then all of a sudden, boom, you're gone." Even after Quinn was fired, he went to say goodbye to Bancel. "No hard feelings," he remembers Bancel saying. "Hit me up in a month for lunch and we'll talk about your career." Quinn did reach out but it never led to anything. "It taught me what kind of businessperson Stéphane is, which is brutal."

When Schrum left, he asked Moderna for the shares from stock options he had exercised. He had been promised in his employment letter that he would get options that would vest quarterly. According to Schrum, he exercised some of his options on his one-year anniversary and the rest before he left the company on his own in April 2012. But Moderna never gave him certificates for the stock he owned or any document showing he had exercised his options. "I went back and asked for them again, they claimed that they didn't have any records that I ever did that," Schrum said. Moderna was claiming that a brilliant scientist trained at Harvard had simply walked away, forgetting to take his company shares with him. Derrick Rossi, who was on the board at the time, estimates that Schrum may have been offered as many as 100,000 shares. If he had exercised even a quarter of those, after stock splits, the loss would be eyepopping—in the tens of millions of dollars. Schrum would not say exactly how many shares he lost but acknowledged they would be worth millions of dollars today. In July 2021, as the stock was selling for $343 a share, Schrum said it is still painful for him to think about and he was contemplating what to do about it.

One scientific advisor said Moderna did the same thing to him, denying him the shares he'd originally been promised.

And Rossi is convinced that Bancel even tried to deprive him of much of his founder shares. One day in 2013 Bancel called Rossi into his office. He told Rossi he had decided not to renew his contract. Rossi said, "Okay, that's fine. But it's not up for another year." But as they continued to talk, Rossi suddenly realized that Bancel was proposing to end his contract immediately. Because his shares in the company vested over four years, that would mean he would lose a quarter of his stake in the company. In 2021, that stake would be worth hundreds of millions of dollars, if not more. "I said, 'No fucking way are you going to do that,' and I started screaming at him in his office. He totally did an about-face. I think he was just expecting that I would roll over. He said something like, 'Oh, no, no, no. Of course your founder's stake is protected.'" To Rossi, it was a sign that Bancel was trying to make his own shares worth as much as possible.

Bancel's seeming obsession with money, however, may be why Moderna would become so successful. "If the goal of a company is to make money and to make products, I can tell you that Stéphane Bancel is laser focused on making money," said an early advisor to Moderna. "In the first instance, probably for himself, but in the second instance to make Moderna a successful company." Weissman and Karikó had tried to start their own company like Moderna but failed to raise enough money. Looking back, Weissman said Bancel's biggest contribution to the advancement of mRNA science was his ability to entice investors. A former insider said, "If you are asking me would you work for him or should he marry your daughter, we would have a different conversation. But if you're asking me as an investor is this person going to drive this company to success, I give him very high marks for that."

By March 2013, Moderna had a website that showed 11 products in the pipeline. The company was not pursuing vaccines then because they did not seem lucrative enough. The products in the works addressed unspecified genetic disorders, blood disorders, and cancers.[35] Biotech startups tend to focus on one or two products, but Moderna always assumed that mRNA therapies were so similar that you could develop many at the same time. Soon, the stealthy company would have a bombshell announcement.

Moderna announced a collaboration worth up to $420 million with pharmaceutical giant AstraZeneca. It was a massive deal for a little-known biotech startup that had as yet produced nothing. The deal included options on up to 40 drugs, including possible treatments for cardiovascular, metabolic, and renal diseases as well as cancer over a five-year span.[36] Luke Timmerman, publisher of the online biotech newsletter *Timmerman Report*, said the deal came like a "bolt out of the blue . . . People looked at this and said, 'Who the hell are these guys? And why are they getting so much money?'"[37]

Much of it certainly had to do with Bancel's sales skills. "He's got a silver tongue," said former Moderna employee Justin Quinn. "He's an extremely gifted salesman." Bancel shows off those skills in a TEDx Talk he gave on a small stage on Beacon Street in Boston in 2013.[38] Wearing a casual gray sport jacket, a plain blue shirt, and no tie, Bancel delivered a 14-minute speech that laid out the advantages of mRNA in simple terms. By the end, audience members must have wondered if Moderna was about to revolutionize medical science. At the core of the talk was work that had started in Rossi's laboratory but had been completed by Moderna cofounder Kenneth Chien at his Harvard laboratory. It explored whether you could regrow dead heart tissue after a heart attack using mRNA coded for the protein vascular endothelial growth factor, or VEGF for short. This is the treatment AstraZeneca has spent years trying to develop. Listen to how Bancel sells the science a few months after the AstraZeneca deal was revealed:

> Let me walk you through one example of an application that has been tried over the last 40 years across many companies that has always failed. How do you regrow the heart after a heart attack? So Dr. Ken Chien, who is one of Moderna's academic cofounders, he is a professor at Karolinska [Institute] in Sweden, has spent his life studying the heart. He is a cardiologist, and what Ken taught me is that on every one of your hearts we have stem cells until the day we die. All your life you have stem cells sleeping on your heart tissue. And what Ken taught me is that there is one protein in the body called VEGF, that if you send that protein to the stem cell it's going to tell that stem cell, go back to work. Make more heart. Okay? And so what Ken has done . . . years ago, he tried to make a drug with recombinant protein—the technology that the biotech industry has used. He made the protein VEGF in *E. coli* in a reactor in a factory, and they injected it into the heart after a heart attack. The problem is when you inject it, it goes into the circulation in your blood way too quickly. And so to have enough VEGF to make the stem cells wake up to make new tissue, you have to go very high in dose. And the problem with going very high in dose as it goes around your body, you have a ton of side effects. So the drug failed in trial. And then 10 years after when gene therapy was discovered, Ken jumped on it again . . . The problem with gene therapy is when he injected it, you cannot stop it. It keeps making VEGF

> over time, and that's not good either. So when Ken saw that technology, mRNA, he said, "That's exactly what I've been looking for for the last 30 years." And he tried, and the first time he tried, he got those results that I'm going to try to walk you through.

A line chart pops up on a big screen on stage. It was actually two charts combined into one from an article Chien's team had just published in *Nature Biotechnology.* Mice had been induced to have heart attacks and then given either a placebo or one of several treatments. There was a line for the mRNA treatment, one for a DNA or gene therapy treatment, and one for the placebo or no treatment. The experiment had in fact tried two DNA treatments, but Bancel only used the one with cataclysmic results: a survival rate of only 20% after 30 days. The other DNA treatment had results nearly identical to the placebo.[39]

> So the blue line is the animals that get no treatment. They just get a water injection, a placebo. And you can see . . . only 15% of the animals still survive one year after a heart attack. In green, you have gene therapy. [The green line falls dramatically in the first few days.] Not so good as a drug. You kill the animal faster than if you do nothing because it just keeps making VEGF and you just have the whole body going crazy . . . The red line is what you see with one dose, one injection of mRNA VEGF within 48 hours after a heart attack. What this one dose is going to do is for 48 hours, it's going to make a lot of VEGF, sending a very strong signal to the stem cell on your heart, saying make heart tissue and then it goes

> away. The mRNA gets degraded after 48 hours. It's gone. But the VEGF is here and it has instructed the stem cell to regrow your heart.

The survival rate for the mice injected with mRNA was impressive: more than 60% after a year. Bancel then showed a slide of a cross section of two hearts from mice after one year. The control, an untreated animal, shows terrible scar tissue on the damaged heart. The one treated with mRNA coded to produce VEGF shows no scar tissue. The heart vessels have been fully restored. It was a mind-blowing chart, a sort of Frankenstein-like regeneration of dead tissue. The audience must have thought they were witnessing a medical miracle. Bancel ends the talk by marveling at the awesome responsibility he has to make sure these medical miracles happen.

> That's the power of this technology. So once you understand you have a technology like this, we have always been paranoid. How do we not mess it up? When you understand that your technology can change society, impacting so many millions of lives, how do you build that company?

Not everyone inside Moderna was convinced that the heart experiment would ultimately work, however. "It is objectively very dramatic to look at. New blood vessels formed around a heart that had had a heart attack," said Jason Schrum. "That being said, recombinant VEGF protein had already been attempted in the clinic for recovery from heart attacks. So the concept of using VEGF to rescue cardiac tissue post heart attack wasn't new at all. The problem with the recombinant protein approach and presumably what I would expect with the messenger RNA is the blood vessels that were formed were completely unstructured. It was just sort of like random blood vessels that were forming over the heart. So that's not how you want to do it."

The experiment led by Chien showed improvement in blood flow within the vessels using mRNA over other treatments and it showed better survival rates, at least in mice.[40] Rossi, who was a coauthor on the study, believes the results were impressive. AstraZeneca would have been interested in that project, given that it was the most advanced research Moderna had in its portfolio at the time. It was another case where the science had been done in an academic lab instead of within Moderna. But Rossi was surprised that it became Moderna's most prominent project so early. Unlike making vaccines, which is relatively easy, regrowing heart tissue in humans is extraordinarily complicated. It is a project that is not likely to pay off for many years.

AstraZeneca has been working on regrowing blood vessels ever since the TEDx Talk. In 2016, they did a phase 1 trial in Europe to test the safety and tolerability using mRNA to generate VEGF. The randomized, double-blind trial enrolled 42 men between the ages of 18 and 65 with mild type 2 diabetes.[41] The shot was given in the forearm, and results showed that VEGF was produced and blood flow near the site of the shot increased for seven days.[42] AstraZeneca went into a phase 2 trial, testing the method on people having bypass surgery.[43]

"I've watched the video, and he just has a way of relating to people and making them feel like they came out of that enlightened," Quinn said. "He's a CEO whose job it is to sell an unproven technology. Like that's any CEOs job in a startup biotech. Their job is to raise money on the unknown. And they just use the data that they can have available to them and try not to step over the ethical boundaries . . . He carefully crafted his presentation so that you feel like you want to buy what he's selling."

Moderna veered into vaccine development not long after that. It was in many ways the most logical first step for mRNA, the step that Weissman had in mind. One problem in the commercial development of vaccines is that most are cheap. "The Gates Foundation says that any vaccine to be used in Third World counties needs to have a maximum cost of $1.50," Weissman said. Moderna announced it had been given $25 million from the government to research and develop mRNA products to generate antibodies for infectious diseases.[44] The dollars came from the Defense Advanced Research Projects Agency, created during the Eisenhower administration during the Cold War. The Soviet Union had just launched Sputnik in 1957, creating fears that the communist adversary had surpassed the United States in missile technology and may be able to drop a hydrogen bomb on the mainland.[45] DARPA rests within the Department of Defense, and its mission includes being prepared for biological warfare.

Within just six years, Moderna had raised $1.9 billion and had a market value of $5 billion, even though they were just a research company with no products. In 2017, a TEDx Talk interviewer said, "They are the Airbnb of biotech. You are looking at the Mark Zuckerberg of biotech." At the time, Moderna had 12 drugs in development, but only 250 human beings had been injected with an mRNA drug in clinical trials.[46]

A key turning point came when Moderna would try to develop a vaccine for an epidemic that was under way, starting in Brazil and reaching the southeast United States. The mosquito-borne virus was called Zika, and the Obama administration made it a high priority to develop a vaccine. Moderna was unsuccessful, but that effort would ultimately change the fortunes of the company. It was that effort that led the National Institutes of Health to partner with Moderna and ultimately to select it to collaborate on a COVID-19 vaccine.

CHAPTER 8

Tackling a Childhood Disease

Around the time Moderna was formed, Barney Graham at the National Institutes of Health was embarking on new research that would become pivotal to the success of the COVID-19 vaccines a decade later. Graham would over time develop a new obsession: stopping the next pandemic. The mission of the Vaccine Research Center at the National Institutes of Health had been laser focused on finding an HIV vaccine. But Graham was interested in other viruses too. In the process, he became the one person who foresaw the potential of a deadly coronavirus pandemic and started preparing for it. Without Graham, we might not have had a vaccine so quickly.

The course he set really began early in his career. Graham grew up in Olathe, Kansas, a suburb of Kansas City. Yet he feels more closely connected to a speck of a town, Paola, where he lived in his senior year of high school. Even though his home was 25 miles away, Graham spent his summers and weekends driving south to work on his father's farm near Paola. There is still a lot of the farm boy in him. Graham is a towering man with a gentle demeanor. He talks deliberately and carefully in a quiet voice. He is humble to a fault. He and his wife hold hands and say grace before meals.

His father was a dentist who Graham says was always in a hurry. He graduated college in three years rather than four. He went to dental school instead of medical school because he thought it would be faster. Graham grew up in a household where speed was of the essence. His father was also frugal. Any time Graham spent money, his father would say, "That's a lot of $5 fillings."

Graham's father and his dental partner bought an 800-acre farm when Graham was a teenager. Eventually, he bought a 40-acre home site nearby as well. Though Graham graduated as the school valedictorian and a sports star, he and his older brother spent countless hours helping the man hired to manage the farm. He learned how to build fences and care for animals. He loved hands-on work. His father tried raising quarter horses but discovered that it required too much expertise. So he switched to cattle—not just any cattle, but massive Santa Gertrudis cows bred by the King Ranch in Texas, the largest ranch in the United States. The cows' hides are the burgundy leather identified with luxury items. But the example Graham likes to use is the King Ranch edition of the Ford F150 pickup truck. One of Graham's jobs, ironically, was to ensure the cows were vaccinated. He also had the unpleasant task of branding the cows with a hot iron. He shrugs this off as just part of farm life. Graham also help breed the 2,500 pigs twice a year. The fields were abundant with rows and rows of soybeans and wheat.

Despite being an excellent student, Graham wasn't fond of reading, writing, or social studies. He preferred math. He managed to get into elite Rice University in Houston, where he majored in mathematics. Something about numbers and patterns appealed to him. There was a beauty in solving problems. But in his junior year, his interest started to wane. He took a class in topology, which he describes as the mathematics of knots. It was giving order to random undulations. Graham found it mind-numbingly theoretical. If this was what mathematicians did, Graham feared his life would lack purpose. He craved a subject more tangible, more hands on. So he switched to studying biology and became a premed student.

At Rice, Graham lived in a two-bedroom suite with three other students. One was Bill Gruber. The two used to toss water balloons from the top of the dorm to kill time.[1] Gruber would later become the head of clinical trials at Pfizer. By chance, the old college roommates would end up competing to see who could make the first mRNA vaccine for COVID-19.

Graham went back to Kansas for medical school. While some are drawn to the medical profession for the money, Graham says that never interested him. He knew that rather than becoming a surgeon or a cardiologist, he wanted to stay in academics and do research. While it was less lucrative, research offered a more stable, less hectic life. He would meet his best friend while at school. In his second year, he was paired up with a classmate to see a patient for the first time. After learning how to use a stethoscope, Graham and Cynthia Turner walked into a room in a Veteran Affairs hospital to examine a patient. The man looked at them both and told Graham, "You're going to have to leave because I only work with lady doctors." Graham started dating Turner. Their first date lasted 14 hours and involved a jazz concert, dinner, a movie, a jazz club and finally breakfast. "We didn't get bored," Turner said. "We just talked. He was very gentle, with a kind spirit."[2] After just three months of dating, the two decided to get married. Turner, who is African American, had three children with Graham, who is white. Anyone who spends any time with Graham learns that he is passionate about civil rights and social justice, topics he brings up often.

During medical school, Graham went to work briefly at the National Institutes of Health. He came back excited about everything he learned. His wife said, "I bet you're going to be back at NIH someday." Graham shook his head, saying, "There's no way. There is no pathway for me to get back to NIH." Graham studied internal medicine. Turner had done a lot of sewing growing up and thought she might become a surgeon. But then she went into pediatric training only to decide two years later to switch to psychiatry. They both ended up doing residencies at Vanderbilt University starting in 1979. While doing his chief residency at Nashville General Hospital, Graham saw one of the first patients in Tennessee with AIDS, which at the time didn't even have an official name. It was temporarily called gay-related immune deficiency. That was in the fall of 1982. The patient had moved back to Nashville after spending most of his adult life in New York, where he caught the virus. He was homeless and penniless. Graham watched for the next four weeks as this man suffered from five or six rare opportunistic diseases at once. It was devastating for Graham to witness. "Trying to understand how that many things that will be going on at the same time in one person was sort of baffling and scary, too. And some of the doctors, especially the pathologists, didn't want to participate in his care or treatment because they were scared. We don't know what we're doing. We don't know what this is. We don't know how anything about it," Graham recalls. It was an experience Graham cannot forget. He still remembers the patient's name and details of all his diagnoses. The next year he saw more AIDS patients as a chief resident at Vanderbilt University Medical Center.

Graham knew that to get the full respect of his colleagues, he should specialize. So he decided to study infectious diseases. He convinced two of the supervisors at Vanderbilt to create a unique fellowship for him on the subject. "If you saw what [AIDS] was doing in the 1980s, you were motivated to become a viral immunologist," he said. "This was threatening our whole way of living." Still, HIV was largely ignored by both the medical community and politicians. Graham sees it as an endemic problem in health care. "This is the way minority people are treated even to this day. There is a dismissiveness and wanting to set it aside . . . That's another privilege white people have. White people don't have to know anything about Black people to survive in this country . . . But at the time, HIV and AIDS was from a political standpoint dismissed as a problem."

At the end of a fellowship, you're supposed to do research. The only group at Vanderbilt working on viruses was a pediatric group. Graham would seek out Peter Wright, who headed pediatrics, to focus on infectious diseases. Naturally, that led him to be interested in contagious childhood diseases. One in particular stood out. He would talk to Wright about respiratory syncytial virus (RSV). Though a mouthful to pronounce, RSV is one of the most common diseases in the world. It afflicts every child by the age of three. Immunity is short-lived, and we all catch the disease again periodically. Usually, it's no worse than a cold. Your child gets a runny nose, coughs a lot, and may have a fever. The symptoms subside within a couple of weeks. But RSV can lead to serious complications, especially in infants and seniors. Each year, it sends nearly 60,000 children to the hospital for an overnight stay and another 500,000 to emergency rooms. Indeed, it is the leading cause of hospitalizations for preschoolers. Although deaths in children are rare, the Centers for Disease Control and

Prevention report that an average of 14,000 adults age 65 or older die annually from RSV.[3] In fact, it rivals influenza each year for cutting the lives of seniors short. RSV rarely gets mentioned as being as deadly as the flu, however. "People just don't know that," Graham said. "Hardly anyone seems to know that. The flu virologists were always much better at publicity, partly because of the 1918 pandemic."

RSV can be a parent's nightmare. Imagine a newborn struggling to breathe, watching the skin on their chest get sucked between their ribs as they exhale, or their head bobbing with each breath, or their skin starting to turn blue from lack of oxygen. This can happen as the inner linings of small airways in the lungs swell, a condition called bronchiolitis. As the airways shrink, the newborn starts wheezing. The airways can also pop with each breath, making a crackling sound. Pneumonia may develop.

Graham's obsession with RSV became mythical for those who knew him. Some of Graham's closest colleagues assume scenes of infants struggling to breathe tugged at Graham's heart when he was a young doctor. It would fit his empathetic nature. But in fact, Graham never treated a child with RSV. Not even once. "I worried about them all the time," he said. "I have treated adults with RSV, but we didn't have any decent treatments." His interest in the disease really began purely as an interest in research. On the other hand, Graham was well aware of the impact of the disease. "Our hospitals fill up every winter with wheezing children, and those children grow up to be wheezing older children," Graham said. RSV sent his sister's son to the hospital when he was still tiny, just a few months old. And even as he got older, asthma during exercise became a real problem for him. "RSV has a lasting effect on children," Graham said. One study showed that adults in their 30s and 40s who had been hospitalized with RSV later develop a nonsmoking form of chronic obstructive pulmonary disease. COPD obstructs oxygen from getting to the lungs. It can lead to heart disease, lung cancer, and other ailments.

Graham's fixation on RSV was interrupted for a while when he decided in 1984 he needed to get a grant. "I realized if you didn't have your own money, you didn't have any freedom in academic medicine," he said. Otherwise, he would have to spend all of his time seeing patients at the hospital or teaching classes. He wrote up a grant application and submitted it. Carole Heilman, who worked under Anthony Fauci at the National Institutes of Health, called Graham and congratulated him; they had agreed to give him the grant. Graham says he cannot remember the amount, but an average grant at the time was about $150,000. But three days later, Heilman called back with bad news. "I'm really sorry to tell you this, Barney, but we're not going to be able to fund your grant this year," she told him. President Ronald Reagan announced in January 1985 that to reduce the deficit, he was cutting the number of NIH grants by 23 percent.[4]

Graham was deflated. Soon, however, the head of his department at Vanderbilt said that the university would fund the grant on its own using money from the general budget. The cash allowed Graham to embark on new research while going to graduate school to study microbiology and immunology. Graham spent six months researching pox viruses but realized that he was still obsessed with RSV. He thought to himself, "You know, what I really, really care about is RSV. I need to finish the RSV project." He would start working on an RSV vaccine in 1985. He would end up writing his dissertation on RSV and getting his PhD in 1991.

Peter Wright had established a world-renowned center for vaccines. Though he had been at Vanderbilt for years, he had trained at the National Institutes of Health with Robert Chanock, who first identified and characterized RSV in 1956. Chanock was head of the respiratory virus section of the NIH's Laboratory of Infectious Disease.[5] He tried to develop a vaccine for RSV, but that effort ended in catastrophe. The human subjects selected in those trials were African American babies from impoverished families, a population Graham felt highly protective of. Graham would spend the rest of his career trying to fix what had gone wrong in that tragic experiment.

CHAPTER 9

Tragic Trial

Robert M. Chanock was a pediatrician who was later hailed as "one of the top 20 virologists in history [who] covered a broad range of infectious diseases with amazing productivity."[1] While he was a resident at the University of Chicago in 1948, the daughter of one of the faculty members there was admitted to the hospital with a severe case of the croup, a disease in which the upper airway swells up, causing the flow of air to be obstructed. Croup has a signature barking cough. Doctors had to do a tracheotomy, cutting a hole in her neck and putting a tube into her trachea to save her life. Later, they had to put her on a ventilator. She survived, but it was touch and go. Chanock started to research what caused croup and quickly realized that no one had identified any of the viruses behind it. The same was true for viral pneumonia and bronchiolitis, the lower respiratory tract diseases that filled up pediatric wards during the winter.[2]

To hone his research skills, Chanock went to work in the laboratory of Albert Sabin in 1950. At the time, Sabin was chief of infectious diseases at a children's hospital affiliated with the University of Cincinnati. While Chanock was his chief resident,

Sabin began his bitter rivalry with Jonas Salk to develop the first polio vaccine. Salk was using dead polio virus. Sabin was trying weakened virus.

"I worked for Albert for two years and it was amazing. We hit it off and I was able to survive. He was very severe and very demanding and the many, many people who came to his laboratory for training left after a few weeks or a few months," Chanock later said. "They just couldn't take it."

Sabin was a mean micromanager who criticized his researchers constantly. Chanock said he survived by not screwing up too much. "One day, I really screwed up terribly, and his response will give you a small insight into Albert Sabin and the great confidence he had in himself. I was very depressed, and he came in my laboratory and put his arm around my shoulder and he said, 'Bob, don't be depressed, you know, I made a mistake once.'"

Chanock decided to study the cause of croup. He managed to isolate the first human parainfluenza virus and later a second one. After he accepted a position at Johns Hopkins, Chanock in 1956 isolated the first human strain of respiratory syncytial virus, "which proved to be the most important single cause of serious viral lower tract respiratory disease in infants and children worldwide."[3]

But soon after that, Chanock was fired from Johns Hopkins when he went to the dean and made serious accusations against his boss. "This man is engaged in fraudulent research. Many of the things he's doing are just not true, not correct. What he has written has come more from his imagination than from the laboratory," Chanock said.[4] He wasn't out of work long. Robert Huebner, who headed the Laboratory of Infectious Diseases at the National Institutes of Health, asked Chanock if he were interested in a job. Chanock had met Huebner before. But he first became aware of him in 1952 when he watched a television dramatization of Huebner's discovery of rickettsialpox on NBC's *Philco Television Playhouse.* It was based on an article from the *New Yorker* on medical detectives.

Huebner was the opposite of Sabin. "He inspired, he energized, and he brought people together, in cooperative endeavors, which were almost unheard of before that time . . . Albert Sabin was the lonely scientific giant. He had very few people in the laboratory. When I arrived at his laboratory in 1950, he had a staff of two; himself and his technician. He knew everything about the various activities in his laboratory. Much of the work he did himself with his own hands. He also analyzed the data himself. In other words, he exercised total control. In contrast, Bob Huebner was one of the great early entrepreneurial scientists, who brought a large group together and trusted each member of the group to do what they were supposed to do. He did this by including controls that allowed him to determine whether bias had entered into the equation or whether everything had been done properly. He had maybe twenty, thirty, forty people working for him and they were all loyal and dedicated."[5]

One of Chanock's pursuits was to find a vaccine for the croup. He teamed up with researchers at Children's Hospital at 13th and V Streets in Washington, DC. The hospital was a massive, old, crumbling building that years later was replaced by a new facility on Michigan Avenue, across the pond from Howard University.[6] Besides Chanock, the researchers included Hyun Wha Kim, an intensely private woman who had escaped from North Korea after the war. She worked at a research laboratory at Children's Hospital. There was also Robert Parrott, the administrator of the hospital. But the marquee name on the study was Chanock.

They conducted their first RSV vaccine trial starting in 1962. The results of that study were never published, but it was mentioned in a later paper. The researchers injected the experimental vaccine into 54 infants less than a year old. It didn't go well. Twenty-one of the babies caught RSV during the next season, and 10 of them had infections so severe that they had to be hospitalized. That's a hospitalization rate of 18.5 percent of all children given the vaccine and nearly 50 percent of the children who caught the virus. A study published in 2009 in the *New England Journal of Medicine* concluded that among children under the age of five, the average annual hospitalization rate for RSV was 3 out of 1,000 children.[7]

Kim wrote, "At the time, we felt that the high incidence of infection and illness in vaccinees was possibly outside the range of normal; however, there was no concurrent control group with which to compare the usual incidence of illness." In other words, they were not willing to blame the vaccine for making the children sicker.

So they tried again in 1965. This time, they injected an experimental vaccine three separate times into 31 babies. The first two injections were a month apart. For the third, they waited three months. The babies ranged in age from two to seven months old. Rather than having a control group of unvaccinated babies, they had two other groups of 40 infants given a trial vaccine for parainfluenza, another common cause of croup.

The experimental RSV vaccine was similar to Salk's polio vaccine. The researchers took a live virus collected from someone's throat and grew it in cell cultures. They killed the virus by putting it in diluted formaldehyde—formalin—and heating it up. "That was at a time when a lot of things like that were working," said Barney Graham, who would carefully dissect the experiment years later. "That's how the flu vaccines were made. That's how a lot of other viral vaccines had been made that worked." The vaccine was tested first in rabbits, guinea pigs, monkeys, and mice. The shots were given to the infants starting in December 1965, the same month Gemini 7 was launched into orbit from Cape Kennedy with two astronauts aboard. It was a time of enormous scientific confidence. Researchers kept giving shots until the following December. RSV swept through the Washington, DC, area from December 1966, when the last shots were given, through April 1967.

But disaster struck. Of the 31 infants in the trial, 20 became infected with RSV. Far worse, 16 of those infants gasping for breath suffered such severe symptoms that they had to be hospitalized. That was 80 percent of the children infected. Kim and the others said they couldn't compare that rate to the general population because the worst cases are usually within the first six months. The children in the experiment were older by the time they were infected. But among the 40 babies given the trial parainfluenza vaccine, only one had to be hospitalized. The infants in the experimental trial spent an average of 10 days in the hospital, four more than usual. But the biggest tragedy was that two children died. One was 14 months old; the other was 16 months. Deaths from RSV among children are exceedingly rare. Researchers halted any more injections of the vaccine on December 29, 1966.

"I can tell you it was a devastating study personally for my father," said Stephen Chanock, who is now director of the cancer epidemiology and genetics at the National Cancer Institute. "I remember several months where my father was very depressed." His father had a very disciplined daily routine. Up at 4 a.m. Reading and writing papers until 7 a.m. Going to his laboratory until dinner. Bedtime by 7 p.m. But routine life at home changed after those toddlers died. Stephen Chanock recalls his mother and grandmother keeping him away from his father for a while because of his distress. As an adult, Chanock worked with his father on RSV research. It was hard though, because he didn't really talk much about the vaccine experiment. It was too painful. Chanock said of his father, "As he would say over and over, 'The world of vaccines is not for people who were weak of stomach.'"

All the researchers in the experiment are now dead. They wrote in their paper that the babies they selected for the experiment "lived at home and came from a population of relatively low socioeconomic status families, primarily Negro." They said that they got the parents' permission. But Stephen Chanock and Barney Graham, who pored over Robert Chanock's work and later got to know him, said the experiment took place at Junior Village,[8] which the *Washington Post* would later call "a concentration camp for homeless children."[9] In fact, under the leadership of Robert Huebner, the chief of the Laboratory of Infectious Diseases, the National Institutes of Health used children at Junior Village as research subjects starting in 1953 and continuing through 1968. The facility was badly overcrowded, making it a petri dish for childhood diseases.

In an oral history, Robert Chanock would talk about the experiments that National Institutes of Health did at Junior Village: "These infants and children were wards of the court because their parents were either in jail, had been killed, were on drugs, or were not able to care for them for other reasons." The experiment was done long before the days of institutional review boards to ensure the protection of human subjects.

Constructed around 1910, Junior Village was initially called the Industrial Home for Colored Boys. Thirteen prison-like redbrick buildings were called "cottages." They were surrounded by a police lot of abandoned cars, a sewage treatment plant, and the city dump. Ninety percent of the children came from impoverished Black families.[10]

Congressman Alfred Santangelo, a Democrat from New York, once stood on the floor of the House of Representative and described touring the institution crammed with 675 children, more the double its capacity. "A visit there will tear your heart out. Children long to be loved, fondled and made part of a family. The District is doing the best it can under the circumstances."[11]

Many of the children were not orphans. Their parents could just no longer care for them. The population exploded because of a federal crackdown on welfare fraud led by Senator Robert Byrd, a Democrat from West Virginia. While Byrd would later change his views and twice become the Senate majority leader, when he worked as a butcher in West Virginia in the 1940s, he organized a chapter of the Ku Klux Klan. Even as a senator, he opposed the Civil Rights Act of 1964, filibustering on the chamber floor for more than 14 hours.[12] Because of the crackdowns, families were denied welfare checks if there was an able-bodied man living at home. The financial stress meant that families could no longer afford to raise their children. Nine hundred children were crammed into cottages meant for a third of that.[13]

In 1971, five years after the clinical trial, the *Washington Post* exposed horrific conditions at Junior Village in a four-part investigative series. A Head Start teacher there, Ruth Hirsch, wrote a 40-page report on conditions there and gave it to the welfare department. That report ended up in the hands of Aaron Latham, a *Washington Post* reporter. His series horrified the city. There were allegations of sexual assault among the children. Staff often sedated children with Thorazine to keep them under control. In

1969, the body of 11-month-old baby was found hung next to her crib, a story that didn't make news at the time.[14]

Nick Robinson, who now teaches English at Claflin University, lived at Junior Village from 1966 to 1972, starting when he was nine years old. He said, "It was a horrific place for me." Sexual assault by a few staff members and other residents was common and rarely punished. When he first arrived, he was horrified to witness in the cottage's bathroom a crowd watching a young boy being forced to perform oral sex on an older boy.[15] His most vivid memory is the fear of going to sleep. "They would turn out the lights and the thing that was really traumatic for me is what would happen after what they call 'lights out.'" The outrage from the *Washington Post* series eventually led to the institution's closure.

The facility offered a controlled environment for NIH researchers, however, and they took advantage of abandoned children crammed together in an environment perfect for the spread of disease. "Junior Village was incredible," Robert Chanock said years later in an oral history. "We studied RSV outbreaks as well as paraflu epidemics and adenovirus epidemics." Children at Junior Village could be monitored constantly in ways that were impossible outside of an institution. "This prospective approach provided incredibly informative data because every infant or child that was studied was his or her own control," Chanock said. "This long-term prospective study yielded a veritable cornucopia of new viruses and new epidemiologic insight for known viruses."[16]

When Robinson read Chanock's account of the experiments at Junior Village, he paused and started to cry. "You know, that kind of distant, analytical voice when they're talking about, you know, babies. It is just chilling and it's disgusting," he said. The children at Junior Village were vulnerable and defenseless.

Robinson remembers being especially horrified by the cottage where the infants were jammed in together. The cribs were packed into an open one-room cottage. He remembers that the babies were constantly sick, with runny noses and a distinctive cough. He also remembers white people spending a lot of time in the infants' cottage. "Those children were always sick anyways. And to know that they were experimenting with those children—that's devastating."

Junior Village was not the only example of scientists using an orphanage to conduct medical experiments. Early vaccine experiments were often done in children. Around the turn of the 20th century, there are several documented cases of experiments conducted in orphanages, despite objections from rights groups. In 1908, three Philadelphia doctors tested a protein for detecting tuberculosis by putting it on the eyes of orphans, causing painful lesions.[17] Researchers at the Massachusetts Institute of Technology in collaboration with Quaker Oats fed radioactive cereal to mentally challenged children boarded at Fernald State School in Waltham, Massachusetts, during the 1940s and 1950s. In a lawsuit filed years later, the victims alleged the experiments were done without consent. The school and company agreed to a $1.85 million settlement without admitting guilt.[18]

When told of the RSV vaccine trial, bioethicist Arthur Caplan at the NYU Grossman School of Medicine said that while standards were different in the 1960s, that doesn't make what the scientists did "any less awful . . . that doesn't let people off the hook for doing obviously horrible things." Caplan contends that even at the time, the RSV vaccine experiment should have raised moral concerns about the choice of human subjects. "I think at the time, people felt research on children should be done on children who didn't quote-unquote count as much. That would be orphans. That would be racial minorities. That would be children in institutions." What's more, the attitude at the time was that children who were being fed and housed by the state "owed us something and that could be repaid by being subjects in study. None of these studies would ever have been done on middle- or upper-class kids."

The vaccine trial in which two children died was just one of at least three RSV vaccine trials done at Junior Village. They started in 1962, 1965, and 1966 respectively. All were complete failures. The last two trials overlapped, with the last one being cut short in December 1966 after only 10 months. In that experiment, NIH researchers gave the same experimental vaccine to 13 infants at Junior Village. This time, they had a control group of unvaccinated children. Chanock was a researcher on that trial too. The lead author on the study was Albert Kapikian, another superstar at the National Institutes of Health. The son of parents who escaped the Armenian genocide,[19] Kapikian was the chief of the epidemiology of the Laboratory of Infectious Diseases under Huebner. In 1998, he became Anthony Fauci's deputy director at the National Institute of Allergy and Infectious Diseases until he died in 2014.[20] Fauci co-authored a tribute to Kapikian in which he wrote: "One of his first scientific efforts was a study of the infections of children in Junior Village, the public orphanage of Washington, DC. He treated the abandoned children like his family and talked about them warmly until the end of his life."[21]

The researchers' findings, published four years later, noted that beginning in 1955, the research laboratory at the National Institutes of Health had a full-time staff of two registered nurses and four practical nurses at Junior Village each day for its studies. There was also a medical staff of the welfare department for the three cottages with the youngest children. In Harrison Cottage, which housed 52 babies, and Arthur Cottage, which housed 45 babies, rectal temperatures were taken daily during the vaccine experiment. Throat swabs were done twice a week. Blood was drawn every three months. The oldest children in Arthur Cottage were excluded from the experiment because they were allowed to participate in outdoor activities and "consequently were not available for close observation."[22]

Although none of the 13 vaccinated children died, the results were disastrous. Sixty-nine percent of the vaccinated children developed pneumonia compared to 9 percent of the unvaccinated children. "The vaccine not only failed to offer protection but also induced an exaggerated, altered clinical response to naturally occurring RS virus infection," the researchers wrote.[23] Having lower tract respiratory infection from RSV as a child is associated with long-term wheezing, asthma, and impaired lung function.[24]

Today, a tragic clinical trial like that would be front-page news and likely career ending for the researchers. But the world didn't learn of the studies until the results were published in two papers in the same edition of the *American Journal of Epidemiology* in 1969. The deaths were not even mentioned in the paper's abstract, a summary of what happened. Instead, you had to read to page 8 of a 13-page article. Even then, the deaths and the trials themselves attracted no media attention. The only real accounts today exist in the sterile language of scientific journal articles.

The results didn't go unnoticed in the vaccine world, however. What the experiment showed was that vaccines made incorrectly could make things worse. It's a condition known today as vaccine-associated enhanced respiratory disease. For some reason, the vaccine made the children sitting ducks for the virus. While research into RSV continued, medical researchers didn't dare to start testing potential RSV vaccines in humans again until 2000.[25] Stephen Chanock said, "That brought to a screeching halt the development of that approach towards vaccination." Peter Hotez, dean of the National School for Tropical Medicine at Baylor University, said, "That really killed RSV vaccines for a generation."[26]

One man who picked up the effort in the 1980s was Barney Graham, who declined to express his personal views on doing the experiments on homeless Black children at Junior Village. The only other work being done to try to develop a vaccine was from a company called Praxis. That effort led to clinical trials in children again, but the vaccine proved ineffective. Graham decided that this would be his mission in life. He wanted to know why those two toddlers died and the other children got so sick. His research would ultimately change the course of vaccine science.

CHAPTER 10

Vaccine Research Center

Donning a graduation gown, President Bill Clinton stood on a stage on a cloudy spring day to deliver the commencement address at Morgan State University, a historically Black university in Baltimore, Maryland. It was May 18, 1997.[1] Notoriously late for everything, Clinton had arrived next to the stage with his motorcade only one minute after the morning ceremony was supposed to begin.[2] Just two days earlier, from the East Room of the White House, Clinton had issued a formal apology to survivors of the Tuskegee syphilis experiment led by the US Public Health Service. It was an obscenely unethical experiment at another historically Black school, Tuskegee Institute. The government enrolled 600 poor Black sharecroppers, the descendants of slaves, in Macon County, Georgia. Of those, 399 suffered from syphilis. Researchers offered the men free health care, but they were lying. Instead, they wanted to know what would happen if you didn't treat syphilis. These men were guinea pigs. The study started in 1932, before penicillin became a standard cure for the disease. But it continued through 1972, even as some of the men went blind or insane from syphilis, when news accounts exposed the experiment and led to public outrage.[3]

Clinton alluded to the horrific experiment in his speech. "Science has no soul of its own. It is up to us to determine whether it will be used as a force for good or evil . . . We must never allow our citizens to be unwitting guinea pigs in scientific experiments that put them at risk without their consent or full knowledge." He then had a plan to restore faith in science and in the nation's public health institutions. "Thirty-six years ago, President Kennedy looked to the heavens and proclaimed that the flag of peace and democracy, not war and tyranny, must be the first to be planted on the moon. He gave us a goal of reaching the moon, and we achieved it—ahead of time . . . So let us today set a new national goal for science in the age of biology. Today, let us commit ourselves to developing an AIDS vaccine within the next decade."[4] The audience applauded politely.[5]

Clinton said an HIV vaccine was inevitable; it was just a matter of time. He announced plans to establish a new Vaccine Research Center at the National Institutes of Health focused solely on an AIDS vaccine. The center would be within Anthony Fauci's National Institute of Allergy and Infectious Diseases. It would be another three years before Building 40 on NIH's Bethesda campus would open its doors and begin its work. Barney Graham would be among Fauci's first hires.

Graham had started working on an HIV vaccine years earlier at Vanderbilt. One day Peter Wright stuck his head in the door of Graham's laboratory and said, "Barney, I need your help with something." Graham was the only person Wright had in his pediatric group who had expertise treating adults. Vanderbilt was starting a new program for an HIV vaccine development. Wright said, "It'll only take five or 10 percent of your time. It won't be a big deal." But then, Wright left on a sabbatical to Geneva, leaving Graham in charge of the program. At the time, scientists were confident that they could develop a vaccine quickly. In fact, when the virus was first identified in 1984, the head of the US Department of Health and Human Services announced that she hoped to have a vaccine ready for clinical trials within two years.[6]

There was a reason for the optimism. Graham was experimenting with a vaccine that was getting a big immune response in cell cultures. In a petri dish, at least, HIV seemed simple to defeat. "Neutralizing the viruses grown in a cell culture is relatively easy. Neutralizing a virus that comes right out of a person is really, really hard. That's especially true for HIV." That continues to be the case today.

The National Institutes of Health created a network of research centers working on AIDS in 1987. It was called the AIDS Vaccine Evaluation Group, and as the head of a program at one of the nation's leading research universities, Graham would be tapped to run the group for two terms. Finally, Graham told the NIH that he couldn't lead the group anymore. "I cannot keep doing this because I have to work on RSV," he said with a chuckle. "I wanted to stay with RSV because it's kind of like your first love. You just don't want to let go of it."

Despite all the time he spent on an HIV vaccine, Graham kept plugging away to figure out what went wrong at Junior Village. "He came to me intrigued about respiratory syncytial virus, which was a virus that I had worked on," said Peter Wright, the pediatrician who had previously worked with Robert Chanock at the National Institutes of Health. "And he came to me fascinated by the enhanced disease that had been seen in early trials on an inactivated vaccine. And he spent a great deal of time in mice trying to figure out why that was." Wright was in favor of trying to understand what had happened. He said to Graham, "Here's a cubicle and there's a vial and there's a freezer. Go figure it out." Graham walked through the same process as Chanock. He grew the virus in cell cultures. He put it in diluted formaldehyde, called formalin, for 72 hours and heated it up to 36 degrees Celsius, which is 98.6 degrees Fahrenheit. The fluids then were centrifuged at 50,000 revolutions per minute. He would then inject the vaccine into mice. He would work on the puzzle for years.

The RSV vaccine was not the only vaccine to show disease enhancement. There was a similar problem with a measles vaccine that was tested and approved for use in the 1960s. With the experimental RSV vaccine, the children got sick during the first winter after getting the shot. But with the measles vaccine, which was also a virus inactivated in formalin, the children didn't have disease enhancement for several seasons of the virus. "Measles enhanced disease, which was called clinically atypical measles, didn't occur until about the fourth or fifth year after you got the vaccine," Graham explained. Measles is much easier to protect against with antibodies. But over the years, the antibodies would wane while the T cell memory wouldn't. As the antibodies became ineffective, the T cells would produce an ineffective antibody against measles. Fifty to 60 percent of children with the vaccine would have a more severe form of measles. They would get a high fever, a rash on their arms and legs, and severe pneumonitis. Many had to be hospitalized. The vaccine was taken off the market for safety reasons in 1967.[7]

Although Graham wanted to continue his RSV research, he was approached by Gary Nabel, the head of the newly created Vaccine Research Center, and Anthony Fauci at the National Institutes of Health. But he now had a good relationship with leaders at NIH. They would tap him to be one of the first six employees of the new Vaccine Research Center. His job was to set up a new clinic for HIV vaccine evaluation. But he agreed to do it only if they would let him continue to study RSV. "People actually didn't really want me working on RSV," he said. Yet, Graham made them put it in writing as terms of his employment. Still, Nabel was willing to let him pursue research on RSV on one condition: that he didn't attempt to develop a vaccine for it. Robert Chanock was still at NIH, and two people in his laboratory, Brian Murphy and Peter Collins, were working on their own RSV vaccine. It was a bit of a turf battle. In fact, Chanock had helped develop a monoclonal antibody to prevent severe RSV in high-risk infants. This group included premature babies or those with congenital heart disease. Injected into the thigh muscle, the drug Synagis offers protection against RSV for 30 days. One vial alone can cost more than $3,500,[8] and the shot must be given monthly. The monoclonal antibodies were developed in collaboration with MedImmune, a biotech company now owned by AstraZeneca.[9] Despite the FDA approval of Synagis in 1998, there was still a need for a vaccine that could offer more lasting protection for everyone. Graham got to know Chanock and those working with him. "My father thought the world of Barney," Stephen Chanock would say. "And really thought he was the future of RSV."

While Graham devoted most of his time to HIV, he continued to do basic research on RSV. That would finally change in 2007, when a board of scientific counselors reviewed his laboratory. In their report, the counselors asked without irony why Graham wasn't working on a vaccine. Chanock had retired by then, and so had his chief researcher, so the politics had changed. Graham gleefully had all eight researchers in his lab start working on an RSV vaccine. "If I hadn't have stood my ground and really persisted in keeping my own lab, RSV would have fallen somewhat by the wayside," he said.

There had been by then five clinical trials for experimental RSV vaccines, two using a part of the virus and three using recombinant protein. All of them had failed.

Graham's quest to develop an RSV vaccine would get an immense boost in 2009 when a 28-year-old postdoctoral researcher working in an adjacent lab on the second floor of the Vaccine Research Center began collaborating with him. Jason McLellan had recently graduated from Johns Hopkins University, where he had studied structural biology. In that specialty, McLellan learned how to make the invisible world of biological molecules visible. Using highly advanced technology, McLellan knew how to make three-dimensional images of proteins millions of times smaller than the head of a pin. Proteins are the essence of life. They are also what viruses use to attack us.

McLellan's life had led him on a trajectory to making vaccines. Born in 1981, McLellan grew up the oldest of three in a modest suburb of Detroit just north of Grosse Pointe. Neither of his parents graduated college. His father managed a Farmer Jack supermarket. His mother handled billing in doctors' offices. The family lived paycheck to paycheck. Yet McLellan was an extraordinarily gifted student. When he graduated valedictorian of his high school, his lowest grade was a single A-minus in French. His parents had nothing saved for college, but McLellan landed a full scholarship at Wayne State University just a few miles from home. He planned to become a premed major but found himself drawn to organic chemistry. He was such a standout that one of his professors asked him to work with graduate students in a lab, a job he split with working in a sporting goods store. It was in the lab that he decided to abandon plans for medical school to focus instead on biochemistry. He graduated summa cum laude with a perfect 4.0 average, helping him to get accepted to Johns Hopkins. As he was graduating, a mentor at the university suggested McLellan reach out to Peter Kwong at the National Institutes of Health, who was using structural biology to try to develop an HIV vaccine. The chance to use his training to relieve suffering appealed to McLellan. He started working for Kwong in June 2008.

But HIV is an incredibly elusive virus. It embeds itself in your genes, where it manages to stay well hidden. Even if a drug could clear all but one genome of HIV, that remnant could later reactivate and make plenty of copies of itself. HIV also mutates constantly inside your body, creating new variants all the time. A person could develop antibodies to prevent infection from almost any known strain, but that hidden gene inside a person can make a new variant that escapes all of those antibodies. HIV is always a step ahead.

Another problem was the nature of vaccines themselves. Vaccines keep you from getting sick but not necessarily from getting infected. After all, they work by creating antibodies ready to pounce on a virus once it enters the body. With HIV, a single virus integrating with your cells has essentially won. You're infected for the rest of your life. Kwong and McLellan were trying to attack HIV by understanding the structure of the virus, but their work seemed to be going nowhere. Was it the fault of the scientific approach or the virus itself? McLellan didn't really know, but he remembers telling his boss one day, "Why don't we pick a different virus where maybe we can see which of these ideas are good or not?" Graham's lab was just down the hallway, and he had heard at a meeting that McLellan was looking for another virus to tackle. Graham suggested RSV. A vaccine for this respiratory illness had its own problematic history. But the ray of hope was that RSV doesn't embed itself in a child's genes.

McLellan knew little about RSV when it was first suggested. In fact, his impression was that Graham's obsession with RSV was grossly underappreciated at the Vaccine Research Center, whose core mission was to defeat AIDS. The public is also largely unaware of the disease. "RSV flies under the radar. The only people I've known who seem to know about RSV are parents who either had their own kid or somebody else's kid hospitalized with RSV," McLellan said. But the more he learned about it, the more he liked the idea of attacking this virus, both for the scientific and humanitarian aspects of it.

Jonas Salk made a polio vaccine by killing the virus and injecting it into people. Salk didn't know what the virus looked like. And in his case, it didn't matter. McLellan's expertise was in a new field of vaccine science called a structure-based approach. It meant understanding the virus at a molecular level. His goal was to make a precise image of the virus and its components and then use that information to plan a counterattack. There were some things other scientists had already figured out.

RSV is highly contagious. It spreads through droplets when someone sneezes or coughs. It can also live on surfaces for hours, spreading if a child touches a contaminated object and then sucks their fingers. The virus attacks human cells with a protein on its surface known as the F protein. That protein attaches to a cell and allows the virus to invade the cell, where it replicates. RSV belongs to a family known as Paramyxoviridae. Previous work had already determined that the F protein of paramyxoviruses came in two shapes. For the sake of ease, these shapes were called by some lollipops and golf tees. The protein looks like a lollipop when it first infects a person. But then, like the creature in the film *Alien*, the protein hurls its tentacles at a cell, ripping it open to allow the virus to penetrate. Researchers called this shape by the more benign nickname, golf tee.

Graham and McLellan reasoned that a successful vaccine should make antibodies that attack the lollipop, known in scientific jargon as the prefusion state. Once the protein unfurls into a golf tee, the postfusion state, antibodies are less effective. So the basic question was this: How can you make a vaccine that replicates the lollipop? McLellan's structure-based approach came down to this: First, figure out what antibodies found in humans who've been infected do the best job of attacking the virus. Next, figure out how those antibodies defeat the virus. And finally, create a protein that would best mimic the shape of the F protein those antibodies neutralize. As simple as it sounds, it would take years of arduous, precise work to find an answer.

To capture the lollipop, McLellan needed to know which antibodies would bind to it but not to the golf tee. It's like finding a butterfly net that would catch only a certain type of butterfly. McLellan already knew that most antibodies bound to both shapes. So the first step was to eliminate any antibodies that would bind to the golf tee. Once this was done, the next step was to figure out which of those antibodies would bind to the lollipop and also neutralize the RSV virus. With the help of collaborators in China about 2,000 antibodies were tested. It took two years of work. In the end, they identified a single antibody, called 5C4, that would attach solely to the lollipop and effectively neutralize the virus. On his own, McLellan one night while watching television decided to search through patents on the United States Patent and Trademark Office website. He was looking for any additional antibodies that only bound to the lollipop. Through a stroke of luck, he found two more.

The next step was to use DNA to send instructions to cells in a flask to grow both RSV's F protein and the antibodies together. The antibody would bind to the protein, locking the F protein in its lollipop shape. Then comes the true magic of structural biology: crystals. The lollipop is far too minute to be seen by the naked eye or even by a regular microscope. In order to get an image of it, it must be crystallized, the same way you turn sugar into rock candy. The proteins collected are put in a special solution. A tiny drop of this mixture is hung over a pool of crystallization solution in a vial. McLellan would have up to 500 of these vials. It can take days or even months, but over time the water evaporates, and the drop goes into the pool, forming crystals. McLellan would look forward each morning to checking the vials to see if there were any crystals. Usually there was nothing. But then one day, almost magically, a crystal would appear. McLellan would get excited every time he discovered a crystal.

At that point, McLellan would have to take a metal rod with a nylon loop and recover the crystal under a microscope. It takes an extremely steady hand. One false move and the delicate crystals falls apart. It was nerve-racking. If he succeeded, he would instantly freeze the crystal with liquid nitrogen and put it in a container called the puck, because it has the shape of a hockey puck. Then McLellan would get on a plane and take the puck to the massive Argonne National Laboratory 25 miles southwest of downtown Chicago.

This is where the magic of X-ray crystallography comes into play. Somehow, through a stroke of brilliance, 22-year-old Lawrence Bragg in 1912 was able to decipher splashes of dots on photographic film when beams of X-rays were shone through a crystal. It was the key to taking the unseen world of atoms and being able to reconstruct that world into three-dimensional models. Waves of light are larger than molecules. But an X-ray beam has waves smaller than subatomic particles. Bragg and his father were able to shine an X-ray beam through a salt crystal and create an exact 3-D model of the atomic structure of sodium chloride. The son reduced the diffraction of X-rays through a crystal to a simple mathematical formula that remains one of the greatest achievements of modern science. Bragg and his father won the Nobel Prize in Physics in 1915, and he remains the youngest recipient of the prize to this day. Since then, 28 scientists have become Nobel laureates with the help of crystallography. One of the most celebrated examples was when Francis Crick and James Watson in 1953 were able to surmise the structure of DNA based on an image taken by Rosalind Franklin.[10]

Recognizing the importance of making the invisible world of atoms visible, the Department of Energy has five research laboratories capable of conducting X-ray crystallography. On the campus of Argonne National Laboratory is the Advanced Photon Source, which has a massive storage ring big enough to encircle Wrigley Field. Inside that ring, electrons wind around magnets at near the speed of light to produce an intense X-ray beam a billion times more powerful than the X-ray machine at a dentist's office.[11]

On at least one occasion, McLellan carried the crystals onboard a flight in the puck. That puck is then loaded into the X-ray crystallography device, where the X-ray hits the crystal and then scatters the X-ray to form an image of black and white dots. The trick then is to take that image and convert it into a 3D model of the protein. It's excruciating work. McLellan's first attempt with the lollipop was a failure. On the second attempt, he shipped the crystals to Argonne and worked remotely. To the best of their recollections, McLellan and Graham were visiting their collaborators in China at Xiamen University when McLellan called up an image of the splashes of dots on his laptop. He swiveled the computer to show Graham that the image had been successful. In the world of incremental advances in science, it was a big moment. It meant they would soon be the first humans to see what the lollipop or prefusion F protein on RSV looked like. It could be the key to defeating the virus with a vaccine. When McLellan was finally able to construct a model of the protein, Graham remembers it looked nothing like he expected.

Graham would buy McLellan gifts after major discoveries. He bought a massive crystal he found in Brazil and had it engraved with "Jason McLellan, PhD. Post-fusion RSV structure, February 2011." He also had models of proteins McLellan had discovered made from a 3D printer. Graham would also host a large Thanksgiving party at his home for people in his lab and a clinical trials group he oversaw. When representatives from the Gates Foundation met with Graham to ask about funding research on RSV, Graham suggested they fund McLellan's work instead.

Graham had realized in his years of studying the catastrophic RSV vaccine trial at Junior Village that by killing the virus in formalin, the F protein on the surface changed its shape. It went from being a lollipop to a golf tee. As a consequence, the children's immune systems produced the wrong antibodies. Children who had already been infected with RSV at some point in their lives were not hurt so much by the experimental vaccines. They weren't helped, but at least the vaccine didn't make things worse. Now it made sense. The vaccines produced the wrong antibodies. They would bind to the virus, but they wouldn't neutralize it. As a result, it set off a complex reaction that led to inflammation inside the children's airways, forcing the children to struggle to catch their breath. The vaccine also weakened those children's immune responses to the virus.

Their first paper on the research, published in *Science* in May 2013, identified the F protein in its prefusion or lollipop state. It was an important paper for structural biologists. But then, using DNA to create that protein in monkeys, they discovered it produced an immune response unlike anything Graham had seen in decades of working on an RSV vaccine. Graham was so impressed with the results that he had no doubt they would lead to a successful RSV vaccine and imagined that might be his legacy. "That's the thing I've been most proud about," Graham would recall later. There were still issues to resolve. The vaccine couldn't be given to babies, but there might be protection by giving it to pregnant women. And of course, the vaccine could be given to the elderly, who were most at risk of dying.

The public relations team at the National Institutes of Health persuaded a local Washington, DC, station to do a news segment on it, featuring Anthony Fauci. "It's one of the more serious viral infections that afflict children. Also elderly people can get infected and can get sick," Fauci said, adding that the animal research made it likely that a vaccine might work in humans. The news anchor closed the segment with a wildly optimistic prediction that there could be a vaccine on the market in two years.[12]

As of late 2021, the research that Graham and McLellan generated is still being tested in humans. They remain confident that there will be a vaccine.

Yet the scientific community didn't really fully appreciate what a big achievement the discovery was. The paper won recognition as one of the major achievements for the year, but Graham said many of his colleagues still didn't understand it. The old-school way of thinking was that it was the 574 amino acids in the F protein that mattered, not the shape of the protein. But the way that protein unfurls and rips open the cell is the key to its success.

"They didn't think structure mattered," Graham said. "Some really good RSV biologists, I bet I explained it to them a dozen times before they finally said, 'Ah.'" Even Anthony Fauci said in an interview that he didn't fully appreciate how a structure-based approach had changed vaccine science until Graham explained it to him once SARS-CoV-2 emerged.

While an RSV vaccine would be a major accomplishment, Graham's boss, John Mascola, the VRC director, expected this discovery to fundamentally change the future of vaccine research. It was a long, long way from Jonas Salk's method of killing a virus and injecting it into children. It was even more precise than the breakthrough of using a benign protein from the virus to produce antibodies that became popular in the 1980s. Now, this new approach allowed vaccine researchers to find the most potent antibody and then, using DNA or mRNA technology, have a person's own cells create the precise protein that would generate that antibody.

The Vaccine Research Center director, Gary Nabel, left the National Institutes of Health just three weeks before McLellan made his major discovery about the structure of the RSV F protein. The center at that point was still laser focused on an HIV vaccine. But Graham had blazed a new path for respiratory illnesses by using the science of the HIV vaccine. In Graham's mind, the RSV vaccine was more than a single vaccine. It was a new paradigm for vaccines. It could be applied to other viruses. He started thinking in terms of how to be prepared to stop a pandemic, how to make a vaccine on the fly.

The new director of the Vaccine Research Center saw it the same way as Graham. His center could be expanded to focus on emerging diseases. In September 2013, he asked Graham to be his deputy director and to focus on stopping pandemics. Graham was still working on the RSV vaccine, which at the time of this book's publication is in phase 3 trials. He set up a program to create a universal influenza vaccine. This was driven by the knowledge that the last catastrophic pandemic was influenza in 1918. Experts believe another deadly flu pandemic is inevitable, and Graham wanted to be prepared. Graham also began thinking about other viruses to try to conquer. At the top of the list was coronaviruses.

CHAPTER 11

MERS

A 60-year-old patient who was having severe trouble breathing was admitted to a private hospital in Saudi Arabia on June 13, 2012. The hospital was in Jeddah, a city situated along the Red Sea that serves as the gateway for visitors to Mecca, just 50 miles to the east. Fortunately, the hospital had a virology diagnostic laboratory on the sixth floor. Ali Mohamed Zaki, an Egyptian immigrant, had created the laboratory years earlier. He collected blood, saliva, and mucus samples and tested the samples for a variety of common respiratory diseases. However, they all came back negative. When the patient died 11 days later, Zaki still didn't have an answer. But he wasn't ready to give up. He kept testing the samples with some help from another laboratory for more possible viruses but kept getting negative results. Finally, he tested it himself for coronavirus, and the results came back positive. He immediately assumed it was SARS, the novel coronavirus that spread from China in 2003 with a fatality rate of 15%. But surprisingly, the PCR test for SARS was negative. He did the PCR test several more times, just to make sure. He could hardly believe he had discovered a novel strain of coronavirus. He also didn't think anyone would believe a small laboratory in a Saudi hospital. And if it were true, he was afraid he might lose his intellectual property rights if another laboratory tested it. Even so, he carefully shipped a small sample of what was left to a Dutch company, which was able to recover the live virus and confirm his findings. This was an entirely novel strain of coronavirus, one that was deadly.

Zaki had to make a bold decision. If he told the world about the novel virus, he knew that he would anger the Saudi government and most likely lose his job. But he also realized he could potentially save lives. So on September 15, he sent an alert message to ProMED-mail, the program for monitoring emerging diseases at the International Society for Infectious Diseases in Brookline, Massachusetts. It appeared online five days later. He was right about his job. The Saudi Ministry of Health accused him of illegally shipping the sample without permission. On September 25, he got on a flight to Cairo, leaving all his belongings behind. In the meantime, another sample from a patient in Qatar was also posted online by ProMED-mail, showing the same novel strain of coronavirus.[1]

The World Health Organization alerted countries of the new virus and began tracking it. Within eight months, it had a name: Middle East respiratory syndrome, or MERS.[2] There was good news and bad news about MERS. The good news was that it was not transmitted very easily. There were documented cases of human-to-human transmission within families and health care settings, but it appeared to be a zoonotic virus. That meant it was transmitted from animals to people. In this case, the source was dromedaries, or one-hump camels, who had been infected by bats carrying the virus. It appeared that getting infected required prolonged contact.[3] By June 2013, the World Health Organization had received reports of 55 confirmed cases of MERS, 40 of those in Saudi Arabia.[4]

The bad news was that MERS was the second deadly outbreak of a novel coronavirus in a decade. The first known case of severe acute respiratory syndrome, or SARS, was traced to Foshan, China, near Hong Kong, in November 2002. On March 12, 2003, the World Health Organization issued a global alert about unusual cases of pneumonia.[5] In late March, the Centers for Disease Control and Prevention issued travel warnings for mainland China and Singapore. By the end of the year, there were 8,096 cases of SARS with 774 deaths. There were an estimated 27 cases in the United States. Scientists traced the virus to bats, who had infected civets, mammals with a catlike body. Fortunately, the disease dissipated on its own.[6]

With two outbreaks of a novel coronavirus, Barney Graham knew it was only a matter of time before there would be another. After his recent success with using a structure-based approach to RSV, Graham was looking for another virus to research. Jason McLellan had been looking to leave the National Institutes of Health in 2012. He interviewed at the University of California San Francisco and Dartmouth, in Hanover, New Hampshire. He eventually landed a position at Dartmouth, where he would run his own laboratory starting in the fall of 2013. Graham wanted to continue to collaborate with McLellan and wanted to find a niche for him where he could have an impact and make a splash in the scientific community. Rather than focus on viruses that were heavily studied, Graham decided to focus on a lesser-known virus. In many ways it was Graham's way of letting McLellan work on something that would give his career the biggest boost. Many other viruses had been well studied. The structure of the coronavirus remained a mystery. McLellan, meanwhile, wanted to keep focusing on RSV, but his former NIH boss, Peter Kwong, let him know that his laboratory would also be working on RSV. "So as a young person starting out with a small lab, I definitely did not want to be competing with a huge lab like Peter's and unlimited resources at the VRC," McLellan recalls. Plus, NIH was willing to fund MERS research since there was an ongoing outbreak.

Another virus Graham considered for McLellan was Nipah. With the wrong strain, both MERS and Nipah had the potential of a cataclysmic pandemic akin to the Black Plague. But MERS—which had a fatality rate of 35%—was spreading at the time. If it had been highly contagious, it could have been the worst pandemic ever. Had MERS come a year later, Graham might have chosen to study Nipah instead. But there are thousands of strains of coronaviruses in bats that could one day make the leap to humans. Yet very few laboratories were doing any research on this family of virus. So it created an opportunity for a big discovery. "We'll try to do something with coronavirus and it won't be so competitive. There's not that many labs working on coronavirus," Graham recalls. "We wanted to get the spike protein structure and stabilize it and see if we could turn it into a vaccine. We wanted to use the RSV playbook." Before McLellan left for Dartmouth, he agreed to work on a coronavirus vaccine with Graham.

NIH did have an emergency MERS vaccine program under way. But it didn't use McLellan's approach. Instead, it relied on DNA to make proteins that were not stabilized as they had been for RSV. The experimental vaccine went into a monkey study before it was abandoned. No human trials were ever done. "I wasn't very happy with it," Graham said. "We needed a better approach."

Graham was not only changing his focus but helping to change the focus of the Vaccine Research Center in 2013. In April, NIH halted a large study of an HIV vaccine. The study had started in 2009 and enrolled 2,504 volunteers, mostly gay men. A safety review board realized that those getting the placebo were getting infected with HIV slightly less often. Although split evenly, 41 volunteers given the vaccine were infected while only 30 of those getting the placebo were HIV positive. Although statistically the difference might have been due to bad luck, it was clear that the vaccine was a failure.[7]

The Vaccine Research Center had been created with the sole mission of finding a vaccine for HIV. President Clinton set a goal of doing it within a decade. But HIV is just too wily. Scientists at the center kept trying and coming up empty. Trying to defeat such a formidable virus had taught them a lot, however. And now the center was ready to pivot to focus on emerging viruses. Under new leadership from John Mascola, the Vaccine Research Center shifted funding to other viruses. Graham focused on respiratory diseases, including trying to develop a universal flu vaccine. But first, there was the coronavirus to dissect. "We just had two new coronaviruses in the last 10 years, SARS and MERS. And our efforts to make vaccines for those were not very sophisticated," Graham said.

A more sophisticated approach was critical, because in some of the early animal tests of SARS vaccines, scientists saw evidence of enhanced disease syndrome, especially in a strain of coronavirus unique to cats. There was also a concern that the vaccines actually made mice and macaques susceptible to disease.[8] It was the same problem that had been seen in the early RSV vaccine efforts. "People were worried that coronavirus could also have an enhanced disease syndrome if you didn't make the vaccine right," he said. Both Graham and McLellan chalk that up to bad studies in which poor proteins were made. Graham was now convinced that the problem with the RSV trial was that when the virus was killed by the diluted formaldehyde, it changed the shape of the F protein. The protein had flipped from looking like a lollipop to a golf tee. As a result, the children's immune system produced the wrong antibodies.

Andrew Ward, a scientist at Scripps Research in La Jolla, California, emailed McLellan one day with a favor to ask. He was a guest editor of a special issue of the scientific journal *Viruses*. He asked McLellan to write something about his work on RSV. Scripps Research is a highly respected nonprofit research laboratory. McLellan knew that Ward's lab had a revolutionary new cryo-electron microscope that approached the resolution of X-ray crystallography. It had been able to use it to do structure-based work on HIV, getting an image of a part of the virus that had eluded all other laboratories. McLellan agreed to write something and added, "Hey, by the way, might you be interested in joining us in a collaboration on the coronavirus? Barney and I are working on this and we're trying to get the structure to do everything we did with RSV and we think we need EM." Ward wrote back that he had a postdoc that was interested in coronaviruses. He had four plans on the email. The fourth one was "follow the structure, make a vaccine, get famous and have a beer."

The process began much like the process with RSV. Graham had an antibody that bound to the spike protein, and McLellan used it to isolate the spike protein on the surface of the MERS virus. That protein is the key to MERS getting inside the cell and replicating itself. Block the spike protein, and you can stop the virus. Ward's lab took images of MERS spike protein but was not getting great results. The protein was fragile and kept falling apart. "We had a really hard time making MERS coronavirus spikes," McLellan recalls.

But then came a stroke of luck. Someone from Graham's lab traveled to the Middle East during the MERS outbreak and came back sick. Thinking it might be MERS, someone suggested taking samples to test. If it was MERS, they could take antibodies from him to use in research. The test for MERS came back negative, however. It was a coronavirus, just not a deadly strain. Scientists call it HKU1, one of four types of endemic coronaviruses that have been circulating forever and is a cause of the common cold. It was still a stroke of luck, however. You can't just order coronavirus antibodies from a laboratory. In theory, most of us have these antibodies in our system because we've been infected with less harmful coronaviruses during our lives. But this was a recent, documented infection. So this person had high antibody levels.

NIH scientists had an idea. Why not try to get an image of the spike protein on the common-cold virus? They in essence began trying to figure out how to make an effective vaccine for the everyday cold. They took blood samples from the person. This time, McLellan was able to isolate spike proteins that didn't fall apart. They sent those to Ward's lab. The postdoc Robert Kirchdoerfer along with other scientists were able to get a high-quality image and piece together the structure of the spike protein. Kirchdoerfer was the first human to ever see what a coronavirus spike protein looks like in its lollipop or prefusion state. This was the key to making an effective vaccine against it. McLellan doesn't remember the exact moment he first saw the image. But the two labs had monthly teleconference meetings where they exchanged images.

Their work was published in 2016 as a letter to the prestigious journal *Nature*.[9] McLellan's lab tried more than 100 different ways to stabilize the spike protein. The goal was to keep the spike in the same shape it was in before it attacked cells—the prefusion state. They would come to realize that the key component of the protein that our immune system uses to develop antibodies is hidden from view once the protein flips into its postfusion state when it is attacking a cell. Finally, McLellan's team came up with the idea of using protein fragments called prolines to pin the spike protein into the lollipop shape. It was an idea inspired from work at Janssen, part of Johnson & Johnson. They had used prolines to stabilize the F protein for an RSV vaccine. As technical and as unglamorous as it sounds, the two-proline substitution of the spike protein was the key to the success of any coronavirus vaccine. The vaccine wouldn't just produce a spike protein; it would produce one that looked identical to the spike protein on the surface of the virus. And the components that trigger the production of the correct antibodies would be exposed to the immune system.

Kizzmekia Corbett arrived in October 2014 and started working with the newly created spike protein in mice. Corbett is a native of quaint, scenic Hillsborough, North Carolina, not far from Research Triangle. At the age of 16, she landed an internship in a laboratory at the University of North Carolina in nearby Chapel Hill. She decided then that she would become a scientist. She studied biology and sociology the University of Maryland, Baltimore County. She had been nominated and was accepted into the Meyerhoff Scholars Program, which promotes minority students in science. She was also in a scholarship program at the National Institutes of Health in Bethesda, Maryland. The scholarship included summer internships at NIH. At age 19, Corbett worked in Graham's lab and continued to do so each summer until she graduated. Graham's lab at the time was working exclusively on RSV. When she left, Graham gave her a book with this written inside: "I am also proud of how you have matured scientifically and professionally, and have every confidence that you will achieve your dreams and be very successful."

Corbett went on to earn a doctorate at the University of North Carolina, where Ralph Baric, one of the top experts in coronaviruses, had a laboratory. But Corbett came back to NIH while she was finishing up her PhD because she wanted to do vaccine research. She checked around the country, but her experience as an NIH scholar gave her a leg up there. She managed to work for Graham again. "He is by far one of my favorite people," she said. She describes him as extremely empathetic, worldly, and smart. "He runs his laboratory like a family," she said. That included supporting each team member and even hosting an annual Thanksgiving dinner for everyone.

Corbett was interested in the coronavirus research, in part because it was so low profile at the time. "I didn't want to work on something that was so high profile at the VRC because it does cause some pressure," she said.

Corbett took charge of the mouse studies for a coronavirus vaccine. She would inject a vaccine with the two-proline substitution into the thigh muscles of mice. Then, after a while, blood would be drawn and tested for the presence of antibodies. She would also put a pseudovirus in the mouse blood, which is safer than using the real virus. Then she would check to see if the antibodies would block the virus from invading the cells. The real breakthrough, as Corbett explains, was that the mutation that McLellan's and Ward's labs came up with didn't just work for MERS; it also worked for SARS. Not only that, but it worked for endemic strains of coronavirus that causes colds. Corbett gave different vaccines to the mice. The ones with the proline substitutions had a much better immune response. In fact, the antibodies were on the order of 80 times greater with the new structure-based approach. "That was a big, big deal," Graham said.

Corbett notes that while she was studying vaccine responses, she technically wasn't trying to create a vaccine. It was more of creating a blueprint for a possible vaccine. While Corbett was working on this, other viruses were breaking out around the globe and capturing Graham's attention. It was time to put his pandemic preparedness ideas to work.

CHAPTER 12

Zika

In the spring of 2015,[1] thousands of Brazilians in Recife, a seaside city with beaches along the Atlantic Ocean, poured into medical clinics, suffering from a mysterious illness. Their symptoms included bumpy rashes, fever, and muscle and joint pain. Whenever a new, unknown disease emerges, alarms go off within the public health community. Until doctors can identify a disease, they cannot really fight it.

Blood samples were sent to a university in Caxias, in the southern state of Maranhão, where baffled researchers tested them for various possible diseases. Sixty percent of the people tested positive for dengue, a common tropical disease spread by mosquitoes. But dengue doesn't usually cause conjunctivitis,[2] also called pinkeye, which many people were reporting. All the patients were negative for rubella, measles, and chikungunya, another mosquito-borne viral disease.[3]

On March 2,[4] Brazilian authorities alerted the World Health Organization about the still unidentified outbreak. This is where the detective work done by disease experts becomes critical. Three weeks later, doctors in the state of Bahia pulled serum samples from patients with dengue-like symptoms seeking emergency treatments. This time, 30 percent[5] of the patients tested positive for something that doctors hadn't looked for previously, a mosquito-borne virus called Zika.

The new diagnosis was a huge relief. Although this was Zika's first appearance in Brazil—and on the South American continent—the virus was considered no big deal. Zika was known to usually cause only mild illness. Brazil's then–health minister Arthur Chioro dismissively told reporters, "Zika doesn't worry us."[6]

Zika had a boring history. Scientists first isolated the virus in 1947[7] in a sentinel rhesus monkey taken from the Zika forest of Uganda. The following year, researchers recovered the virus from an *Aedes africanus* mosquito captured from the forest's trees.

The first human infections with Zika were detected in Uganda and Tanzania in 1952. But for the next five decades, the virus remained largely obscure as it sporadically hopscotched mostly eastward from Africa to the Pacific islands. The first major outbreak of Zika in humans didn't strike until 2007, on the Micronesian island of Yap.[8] The closest that Zika had come to Brazil previously was in 2014, when the virus landed in the Western Hemisphere on Easter Island, 2,200 miles off the coast of Chile.[9] There is no treatment for Zika, although infected people produce neutralizing antibodies and their T cells mount an immune defense.[10]

But by the fall, things had changed. Reports began trickling in of high numbers of women giving birth to babies with unusually small heads, called microcephaly. The condition prevents the brain from growing normally. It can cause intellectual disabilities, seizures, facial deformities, and sloped heads.

Regina Coeli Ferreira Ramos at the Oswaldo Cruz Hospital in Recife treated 13 babies with microcephaly in one week. She was used to seeing maybe 12 cases a year. "We haven't seen anything like this before," she told a documentary film crew from PBS who happened to be in Brazil as the outbreak occurred.[11]

An X-ray of one affected infant's head showed empty spaces and calcium deposits in his cranium where there normally would be brain tissue. The baby also seemed to have problems with his vision and hearing.

"When I saw the first baby, the second baby and the third baby, I said, I never saw anything like this in my life," Celina Maria Turchi, one of Brazil's leading epidemiologists, who raced to link Zika to microcephaly, told PBS in a documentary called *Spillover*.[12]

Researchers checked the mothers for syphilis, rubella, and the parasitic disease toxoplasmosis—all known causes of microcephaly. Doctors also screened some newborns for genetic mutations such as Down syndrome. But the search was confusing and inconclusive. Some of the pregnant women had or were previously infected with Zika. But health authorities couldn't say for sure that the virus was to blame for their babies' brain damage.

In October, Brazilian officials acknowledged the suspected link between microcephaly and Zika. Then, a possible breakthrough. Adriana Melo, an obstetrician, in early November found the Zika virus in the amniotic fluid from two of her pregnant patients in Campina Grande, a city north of Recife. That was the first link between Zika and microcephaly.[13]

"At last we had a road to follow," Turchi said. "A map."

In all, the number of microcephalic babies born in Brazil in 2015 shot up by almost 20 times compared to the previous year.[14]

By then, Zika had already begun spreading explosively from its epicenter in Brazil[15] throughout the Americas. First Colombia, then Suriname, El Salvador, Paraguay, and Mexico began reporting locally acquired cases of Zika. Then on the second-to-last day of 2015, authorities recorded the first local transmission of Zika in US territory, in Puerto Rico.

The panic over Zika soon went global. Florida, specifically Miami-Dade County, was a prime target for outbreaks. Miami is the only area[16] of the state with a year-round population of *Aedes aegypti* mosquito, one of the vectors for Zika that Tom Frieden, then the director of the Centers for Disease Control and Prevention, described as the "cockroach of mosquitoes" because it's hard to kill and hard to evade.[17]

This was exactly the type of outbreak that Barney Graham was primed to stop. It was, by then, his job. He was getting slammed with a series of outbreaks: MERS, chikungunya, Ebola, and now Zika. He had tried to create a vaccine quickly for Ebola in 2014 after six fatal cases of diarrhea were reported in Guinea and then spread to the nation's capital, Conakry.[18] By May, it looked like the outbreak may be out of control. Graham managed to get an experimental vaccine into a phase 1 trial and then a phase 3 trial in Liberia. The virus killed 11,000 people, but its spread was waning by the time Graham could finish testing a vaccine. He still believes that had he done the trial in Sierra Leone or Guinea, he may have succeeded. But as it was, it was too late to get results from the phase 3 trial. "That lesson taught us that we have to go fast," Graham said.

Now he had Zika virus spreading throughout South America and posing a threat on US soil. Graham was ready to go at lightning speed to create a vaccine. In January 2016, Graham began doing test tube and animal studies for Zika to identify the best vaccine candidate. No drug company was involved. His aim was to develop a vaccine in time to stop the outbreak.

Soon, doctors began[19] advising women to avoid pregnancy and to put off travel to Zika-affected regions. Flight bookings to Central and South America plummeted. President Obama asked Congress for $1.9 billion[20] in emergency funding. In a guest essay in the *New York Times*, Peter J. Hotez, a pediatrician and microbiologist at Texas Children's Hospital, warned that Zika was a "potentially devastating health crisis headed for our region," one that may turn into "a catastrophe to rival Hurricane Katrina."

By the time Hotez wrote those words in April 2016,[21] the evidence had grown strong that microcephaly resulted from pregnant women transmitting the Zika virus to their fetuses.

Officially, the first cases of locally transmitted Zika in continental United States were detected in July, in the Miami area.[22] Unofficially, the virus probably had been circulating under the radar in the US for months. But the growing alarm failed to spur action from Congress, which left for its two-month summer break without approving Obama's emergency funding.[23]

By April, Graham had sent the Vaccine Research Center's pilot plant for making DNA vaccines the genetic code they needed. "Because we had done a West Nile virus approach with DNA back in the mid-2000s, we thought we could rapidly make a DNA vaccine for Zika," Graham said. DNA vaccines, like mRNA vaccines, use genetic code to teach cells to produce the foreign invading antigen that would produce antibodies. Graham was using the DNA approach because of its speed. Rather than growing proteins in a vat, he could make a vaccine with just genetic code.

Anthony Fauci appeared on cable news to reassure Americans that Zika could be stopped with mosquito control outside of the Gulf Coast states. He didn't believe there would be a major outbreak throughout the United States. When asked on Fox News about the prospects of a vaccine, Fauci said, "The vaccine is more of a longer-term solution . . . Vaccine trials take a long time to prove efficacy or not. So we don't expect to have an answer before early to mid-2018." Fauci's timetable of two years away was not nearly as rapid as Graham's.[24]

President Obama held a press briefing in May 2016 to complain about Congress not delivering on the money needed to fight Zika. The House was offering the administration money previously earmarked for Ebola research, rechanneling it for use on Zika. The Senate was ignoring Obama's $1.9 billion request. Sitting in the White House next to Vice President Joe Biden, with the sound of news cameras whirling, Obama said, "You don't get a vaccine overnight. Initially, you have to test it to make sure that any potential vaccine is safe. Then you have to test to make sure that it's effective. You have to conduct trials where you're testing it on a large enough bunch of people that you can make scientific determinations that it's effective. So we've got to get moving."[25]

Graham was moving. Just 100 days after Graham had sent the sequence to the NIH plant, he had the vaccine in a phase 1 clinical trial with 80 volunteers split between Emory University in Atlanta, the University of Maryland, and the National Institutes of Health.[26] Graham had in essence broken the world's record for getting a new vaccine into human testing. The previous record of five months was set in 2009 for the influenza pandemic.

Researchers also explored if genetically modified or lab-sterilized mosquitoes could stop Zika transmission. Several vaccine candidates showed promise.[27] They include Graham's DNA-based vaccine, a live but weakened chimeric virus made from genes of dengue and other viruses, and experimental mRNA vaccines being pursued in a federal partnership with GlaxoSmithKline, UPenn, and Moderna.[28]

Moderna at this point was not known for vaccines. But scientists there went to work to develop an mRNA vaccine for Zika. Rick Bright, the director of the Biomedical Advanced Research and Development Authority—BARDA for short—decided to throw resources at several different initiatives to make a Zika vaccine. In January 2017, he announced that BARDA would support an effort by Moderna. "We don't yet know which vaccine strategy might prove to be most successful, but the history of drug development teaches us that we need multiple shots on goal if we want to beat Zika," Bright said in a statement.[29]

A year earlier, there were no Zika vaccines in development. Now there were seven vaccines that made it to phase 1 trials. BARDA's goal was to have a vaccine by 2018.[30]

Graham managed to get his vaccine into a phase 2 trial, where he could see whether it would work in humans. The first dose was given on March 28, 2017. Graham had 2,428 volunteers at 17 locations in both the United States and South America.[31] But he wasn't able to go fast enough to get results. The Zika virus began fizzling out on its own. "We just missed getting an answer in this 2,400-person trial that we almost killed ourselves over, because by the time we got it into enough people, the spread of the disease had subsided," Graham said.

Graham learned another critical lesson from Zika. "Despite how fast we had been going, it triggered the idea for me at least that we have got to completely rethink our being prepared, because no matter what we're doing, no matter how fast it is, it's not good enough," he said. Graham began discussing pandemic preparedness on committees at the World Health Organization, which crafted a blueprint for epidemic preparedness. It included MERS and SARS. These near misses also inspired the creation of the Coalition for Epidemic Preparedness Innovations, a private-public partnership. CEPI was launched in Davos in the summer of 2016 with backing from the Bill and Melinda Gates Foundation and the Wellcome Trust.[32]

CEPI came together remarkably quickly after Jeremy Farrar of the UK's Wellcome Trust co-authored a commentary for the *New England Journal of Medicine* in July 2015 calling for a global vaccine-development fund.[33] The article identified 47 diseases and infections for which there was not yet a fully effective vaccine. The list included a universal influenza vaccine to stop another 1918 flu pandemic as well as cancer. MERS and SARS were also on the list. CEPI's mission is to fund the development of vaccines and to stockpile investigational vaccines just in case of an outbreak. Part of the idea is to allow for vaccines to be developed on the fly when an outbreak occurs.

Graham was spearheading the pandemic preparedness approach at the National Institutes of Health. The coronavirus was a perfect example of being prepared for future pandemics. His team at NIH, along with Jason McLellan at Dartmouth and Andrew Ward at Scripps Research, had just developed a substitution of a spike protein that worked against a variety of coronaviruses, namely MERS, SARS, and an endemic strain. This set the stage for NIH to be ready the next time there was a novel outbreak of a coronavirus.

It was not clear, though, how widely accepted the notion of pandemic preparedness was. Jason McLellan submitted the critical paper on how to defeat the coronavirus to several scientific journals, including *Nature* and *Science*. The lead writers were Jesper Pallesen at Scripps, Nianshuang Wang, at Dartmouth and Kizzmekia Corbett at NIH.[34] To some, it seemed like groundbreaking work.

But the article was rejected by five different journals. McLellan concedes the paper was complicated and presented a lot of different findings at once. The reviewers questioned whether the two-proline substitution was clouding some of the findings. Eventually, the paper was accepted by the *Proceedings of the National Academy of Sciences* and published in August 2017. It offered the formula for making a novel coronavirus vaccine. It was the epitome of pandemic preparedness. Soon, most of the authors would jointly submit a patent for the two-proline substitution.[35]

There was one other eye-opening lesson from the Zika vaccine. One other vaccine had made it to a second phase of testing in humans. It was an mRNA vaccine from Moderna. While Moderna's vaccine used genetic code to produce the antigen, just like Graham's vaccine, Graham was impressed that Moderna's vaccine was much more potent. It was evidence that mRNA might be the perfect platform for a vaccine speed drill. Graham had first learned about Moderna in 2013, before the company was even testing vaccines. The company's mRNA approach was like the DNA vaccines Graham was experimenting with at the time. Both used genetic code to make proteins. Shortly after Moderna started its own vaccine program, Kizzmekia Corbett began working with scientists there on a potential MERS vaccine. Graham was still obsessed with the possibility of vaccine-associated disease enhancement. In short, he wanted at all costs to avoid a repeat of the disastrous RSV vaccine trial, a vaccine that could make things worse. There was a history of this in animal studies. But the approach Corbett was taking was entirely different from the vaccine used on those babies in the 1960s. Rather than killing a virus with diluted formaldehyde, she was using genetic code only to create a harmless protein on the surface of the virus. What's more, because of the work of

McLellan and Ward, Corbett was able to produce the proper antibodies to defeat the virus before it attacked cells. McLellan believed that disease enhancement in animal studies could simply reveal a shoddy experimental vaccine.

Graham deliberately tried to replicate disease enhancement in mouse studies. You cannot wait forever for the antibodies to wane over time like they would in humans. So Corbett would give the mice low doses of vaccine that wouldn't produce enough antibodies to fully protect the mice. If there were enhanced disease, the mouse would have rapid weight loss and death when exposed to the virus. But that didn't happen. "We tried to recreate the vaccine enhanced illness pathology syndrome in mice," Graham said. "And we have gotten it, but you have to work really hard to get it to happen." The studies showed that mice given the two-proline substitution consistently did much better than mice given a vaccine without the modification.

Graham was confident that mRNA was a superior platform. One of its advantages was that mRNA vaccines deliver the genetic code to the cells with the use of lipid nanoparticles. In essence, the mRNA is inside a fat bubble that penetrates cells to deliver the mRNA. Some other vaccines use a harmless virus, called an adenovirus, to deliver the vaccine to the cell. It's a standard technique that would be used in the Johnson & Johnson and AstraZeneca COVID-19 vaccines. The virus is altered so you cannot get sick from it. It also doesn't replicate itself inside your cells. And it doesn't alter your DNA.[36] But there is hypothetically a problem with this approach. Over time, humans develop an immune response to the adenovirus. It is seen as a foreign invader. So future booster shots may not work. Graham had seen this problem with an HIV vaccine. It used a virus known as adenovirus 5, which humans had encountered before. So when the vaccine was given, scientists believe there was an immune reaction to the adenovirus itself.

Dan Barouch, a researcher at Harvard Medical School and Beth Israel Deaconess Medical Center in Boston, told CNN in 2021 that he has seen no evidence after testing booster shots that the adenovirus used in the Johnson & Johnson vaccine has invoked an immune response to the vector used: adenovirus 26. "There was a theoretical concern that the generation of anti-vector antibodies by the first shot could impede the use of it again," Barouch said. "I think these data put that to rest."[37]

Still, Graham considered it an advantage that the mRNA vaccine uses lipid nanoparticles to deliver the genetic code to the cell as opposed to a virus. Scientists say it's still unclear whether the lipid nanoparticles could create an immune response if regular booster shots are needed. But Graham suspects that any antibodies against the lipids would dissipate after a few months.

By 2019, Graham was a diehard proponent of mRNA vaccines using a structure-based approach. He was the last author on a report in the scientific journal *Science*. The stage was set for Moderna to be a key player the next time a major epidemic broke out.

It would be a concept that Anthony Fauci himself acknowledges he didn't fully appreciate until his fateful meeting with Graham and Mascola just a day or so before Graham and his associates, McLellan and Corbett, designed the vaccine for COVID-19 on January 11, 2020. When asked when he realized that vaccines could stop a pandemic, Fauci said, "It was right around the time when Barney was showing me the data about the stability of the spike protein and the fact that they were very quickly able to integrate it into the mRNA platform. As soon as I saw that, and obviously he was very successful in doing that, you never can predict 100%, but I said to myself, 'You know, this really looks good. I think we're really going to get there.'" That's why he was confident predicting after the meeting that we could have a vaccine in 12 to 18 months, a prediction that caused jaws to drop even among the most knowledgeable vaccine experts. A small handful of scientists had quietly but radically improved vaccine science just in time for the arrival of SARS-CoV-2.

CHAPTER 13

The Race

Eight days after Graham's team crafted a new vaccine, SARS-CoV-1 first emerged in the United States. A 35-year-old man who lived in the Seattle suburb of Everett had just returned from Wuhan when he began coughing and throwing up. He'd never shopped at the exotic seafood market, however—thought by some to be ground zero for the outbreak, a theory now largely dismissed. On January 19, after four days of symptoms, he called a local urgent care clinic for help.

Despite running a batch of tests, the clinic came up empty. It had no diagnosis. It shipped the man's nasal swab to the Centers for Disease Control and Prevention in Atlanta overnight, gave him an N95 mask, and told him to quarantine at home. The next day, the man got a call. The test results had come back: He was the first patient in the United States confirmed to be infected with the new virus. Public health officials would realize weeks later that there had been an even earlier case in Sacramento.[1]

The urgent care center was run by the same company that managed a chain of hospitals. At the CDC's request, Providence Regional Medical Center went into Ebola mode—a protocol for which they had just finished training. It sent emergency medical technicians to pick up the infected man. The techs wore hazmat suits. They used a gurney equipped with an Iso-Pod, a plastic bubble to keep the virus from spreading.

At the hospital, they wheeled him to a biocontainment unit, with a special airflow system to prevent the virus from escaping his sealed room. Doctors examined the man via a video camera. The patient didn't feel too sick—at first. He took his own temperature and barely had a fever. Later he would develop pneumonia, from which he would recover.

Local public health officials scrambled to track down at least 50 people who'd had contact with the man in the five days since his return from China. They never found anyone sick, leaving them unsure how contagious this new disease might be. The *Seattle Times* quoted health experts saying not to worry too much; coronaviruses aren't usually that contagious until someone has symptoms.

From Davos, Switzerland, a CNBC correspondent asked President Donald Trump on January 22 if he was concerned about a pandemic. "No, we're not at all. We have it totally under control. It's one person coming in from China," Trump said, referring to the patient in Seattle. "It's going to be just fine."

Moderna by this point had already been making the COVID-19 vaccine for the NIH to use in animal studies and clinical trials. The company was now experienced with vaccines, having tested several of them for years. Their development pipeline includes vaccines for the flu, Zika, and RSV. A few had even made it to human trials.[2] Moderna executives thought that their first big blockbuster vaccine would be for CMV, short for cytomegalovirus, a type of herpes virus. While CMV symptoms are usually mild, it can cause serious complications for those with weakened immune systems. It infects up to 80% of Americans. At its first conference call with investors in 2020, even as the pandemic was headline news, Moderna continued to focus primarily on their CMV vaccine. Its development was still years away, but the company expected CMV vaccine sales to top $2 billion a year, perhaps as much as $5 billion. Tal Zaks, the chief scientific officer, mentioned the COVID-19 vaccine only once.[3]

Moderna's experience with vaccines turned out to be critical. It took the company no time at all to begin production of the COVID-19 vaccine. Graham called them on a Monday to give them the genetic code, and the next day production began. It was a quintessential illustration of mRNA's plug-and-play capabilities and why Graham chose mRNA for a vaccine.

Now, even before the animal studies began, the National Institutes of Health needed to find sites to test humans. NIH had for many years had a select number of locations that handled clinical trials for vaccines. In 2019, it called this group the Infectious Diseases Clinical Research Consortium, made up of 10 of the nation's leading academic centers for infectious disease research. NIH put out a notice to the 10 centers in the consortium, asking them to get ready to test the experimental concoction on humans.

On January 27, just five weeks after the first known case of COVID-19 had been detected, Lisa Jackson at Kaiser Permanente Washington in Seattle learned that she would oversee that trial. The trial itself was not all that unusual. The original plan was to test 45 volunteers, healthy adults between the ages of 18 to 55. They would be divided in three groups of 15 to get different doses ranging from 25 milligrams at the low end to 250 milligrams at the high end. No one with a compromised immune system was eligible. Two shots would be given 28 days apart, with months of follow-up. The point was to see whether the vaccine had side effects, begin to settle on its proper dose, and determine whether it was producing neutralizing antibodies to stop a virus.

There had never been a vaccine created in less than four years. The fastest-developed vaccine so far was for the mumps, and even that was a real outlier. At the time, it still was not clear how serious the novel coronavirus would turn out to be. Yet, there was a clear urgency with the trial. Jackson dropped all her other work to focus on the vaccine.

On January 28, Trump was briefed in the Oval Office about the outbreak. An account of this comes from Bob Woodward's book *Rage*. "This will be the biggest national security threat you face in your presidency," national security advisor Robert C. O'Brien told Trump. "This is going to be the roughest thing you face." Ten days later, Trump confided to Woodward that COVID-19 was much more lethal than he was acknowledging publicly. "You just breathe the air and that's how it's passed . . . This is deadly stuff."[4]

Kizzmekia Corbett started injecting mice with the new vaccine at her laboratory in Bethesda on February 4. About two weeks later, NPR correspondent Rob Stein visited Corbett in the laboratory as she explained the process. She pulled open a freezer drawer. "So this is a minus-80 freezer. And we have all of our beautiful coronavirus proteins," she said. They had taken blood from the vaccinated mice and run two tests. Corbett was expecting the results back that day. "Today is actually a really exciting day for you to be here," she told the NPR correspondent. He asked her why getting the mice's blood results back was exciting. "Because this will be kind of the first indication of whether we are eliciting the types of immune responses that we had planned to . . . Whether it might work." Graham still wasn't sure just how serious the epidemic would get. He told NPR, "If it becomes a virus that comes back year after year, like the flu does, then the vaccine would be a useful thing to have to reduce the amount of disease caused by the virus."[5]

Stein had left before results came back. They had run two tests. One was to mix the spike protein with the blood of the vaccinated mice to see if it would produce antibodies. The next test was to mix a pseudovirus—a safer approximation of the actual virus—with the blood and see if the antibodies would block the virus from getting into cells. Running tests with the real virus requires extraordinary safety precautions. Peter Wright happened to be in town to attend a scientific meeting, staying at Graham's home overnight. Wright was Graham's old colleague from Vanderbilt who had helped him start his RSV vaccine research. He was now at Dartmouth. After the meeting, around 7 p.m., Wright and Graham drove to the VRC campus in Bethesda to check in on the mice study results. "I remember quite vividly going to the lab with him that evening and seeing the first results in mice that the vaccine might be immunogenic—might induce antibodies," Wright said. Graham was visibly excited. One of the tests is very easy to read because it changes color if the antibodies are present. They were. "That was a big moment," Corbett said, adding that it was not yet time to get too excited. "It's hard to have had celebrated then, because you were really running a race and time was of the essence."

For Graham, the mouse results were simply reaffirmation that he was right to be so sure of the vaccine, confidence earned by devoting so much of his life to trying to develop a vaccine for RSV. Getting good results from the mouse study cleared the way for the first human trial to begin. A process that typically takes months—approving a study design—became Jackson's sole priority and was completed in days.

At Kaiser Permanente, Jackson had not even begun hiring nurses and getting supplies ready when bad news struck at a suburban hospital on the other side of Lake Washington from her downtown research office. Quite a few patients at Kirkland's EvergreenHealth Hospital had pneumonia. This by itself wasn't unusual. But the medical director of infectious disease, Francis Riedo, decided to test two of them for coronavirus. The CDC had just dramatically loosened the criteria for testing that day, February 27. To get tested, a person no longer had to have traveled to China or have had contact with someone with a confirmed diagnosis. The new guidelines gave doctors the ability to use their discretion to test people who were seriously ill and had COVID-19 symptoms.[6] In Riedo's mind, this was a longshot. He fully expected the results to come back negative. But he went ahead with it anyway, just in case. He picked two patients for an arbitrary reason: both happened to be in hospital rooms that could easily be sealed off if the results came back positive.

Swabs were collected that night and sent to the local public health office. The next morning, Riedo was shocked by the results: Both patients had COVID-19. One—a man in his 50s who had never left the country—died that same day. The other, who came from a nearby nursing home, died later. The two patients had no connection.

This was catastrophic news. Riedo realized that the virus had been circulating in the Seattle area for weeks. The statistical probability that two people with no connection to China or to each other both had COVID-19 was that the virus was already out of control, past the stage of contact tracing to limit the spread. Soon, the rest of the nation would come to the same realization.

Two residents of a nearby nursing home, Life Care Center of Kirkland, also tested positive. Fifty more were awaiting results.

At a standing-room-only press conference the next day, Riedo predicted: "What we're seeing is the tip of the iceberg."

Overnight, Seattle became the nation's first hot spot. The CDC flew in experts. A local school district decided to shut down. Governor Jay Inslee declared a public health emergency. Within a few days, much of the rest of the country would go into a lockdown too.

This created a serious problem for the clinical trial at Kaiser Permanente. By pure chance, it was being done in the middle of the nation's first hot spot for COVID-19. If the volunteers or the staff at Kaiser Permanente got infected, it could jeopardize the experiment. Officials at NIH realized they needed to pick a second site quickly. Researchers at Emory University in Atlanta were told to ramp up for the clinical trial.

Although Moderna had a head start because of its alliance with Barney Graham, major pharmaceutical companies with vast experience in making vaccines were not far behind. Pfizer quickly became the most formidable competitor. One of the scientists leading that effort was Graham's former college roommate at Rice, Bill Gruber. Although Graham was collaborating with a competitor, he shared advice freely with Gruber.

Pfizer sized up the options and decided on its own to collaborate with the German biotech company BioNTech to use the mRNA platform. Katalin Karikó had been finally released by the University of Pennsylvania despite having through sheer persistence made what in hindsight was a groundbreaking scientific discovery in medicine. Karikó courted a job offer from Moderna in 2013, but she really didn't want to work there. So she leveraged the offer in negotiations with BioNTech. Her biggest concern was that they might not agree to use modified mRNA, although they reluctantly did, according to Karikó. In hindsight, she says he's happy she got fired from the University of Pennsylvania. "There wouldn't be no Pfizer/BioNTech vaccine if he was not showing me the door." Karikó and a colleague set BioNTech on a path that led to its ability to quickly produce a vaccine.

BioNTech and Pfizer decided to test a range of possible vaccine candidates, but the one that they ultimately picked was virtually identical to the Moderna vaccine. It would use the two-proline substitution of the spike protein that Jason McLellan had designed. That approach was outlined in a scientific journal article published in 2017, so it was available for any vaccine maker to copy. The design was patented by the Department of Health and Human Services, which listed McLellan, Graham, and Corbett, among others, as inventors. The government and the scientists would all potentially have a stake in royalties from the vaccines. Johnson & Johnson also used the two-proline substitution even though it relied on a completely different approach. The J&J vaccine encapsulated the genetic code delivered to the cell in a harmless virus, called an adenovirus. Viruses are expert at invading cells, but adenoviruses don't replicate or cause any disease. Of the major vaccine makers, only AstraZeneca chose not to use the two-proline substitution. "As politically correct as I can say it," Corbett said of the other major vaccine makers, "they're catching up because they are copying and pasting."

On March 2, President Trump brought in several key executives of the vaccine and drug makers to the White House to meet before television cameras with the newly formed White House Coronavirus Task Force. Alex Azar, the secretary of Health and Human Services, started off by saying: "The president will be asking you: How can we make it faster? How can we make anything faster? How can we challenge some of those normal pharma timelines?" All the major players were there, including Pfizer and Johnson & Johnson.

But the first vaccine maker to speak was Dan Menichella, CEO of CureVac, the German company that had been working with unmodified mRNA vaccines since 2006. CureVac had still not produced a product. "Our technology platform is fast and it's agile," he said. He expected to be in a phase 1 clinical trial in three months. Executives from Sanofi and GlaxoSmithKline, who would collaborate on a vaccine, were there. They acknowledged that because they were using proteins, which take time to produce, it would be a year before they would be in clinical trials. Leonard Schleifer, the head of Regeneron, said he hoped to begin producing monoclonal antibodies for treatment of COVID-19 as soon as August. Then came Stéphane Bancel, CEO of Moderna: "We were able to go so fast because we are working, for many years, with the NIH and we had worked with Dr. Fauci's team on the MERS vaccine for the Middle East respiratory syndrome, which is a coronavirus. And so we're able to move very, very fast from a few phone calls to getting a vaccine made ready for the clinic."

As Azar had predicted, Trump pressed Bancel on how quickly he could have a vaccine ready. Bancel avoided the question, saying he hoped their vaccine would be in a phase 1 trial very soon and then a phase 2 trial a few months from now.

"So, you're talking over the next few months, you think you could have a vaccine?" Trump asked.

"Correct. Correct. With phase 2," Bancel answered.

Fauci jumped in to clarify: "You won't have a vaccine. You'll have a vaccine to go into testing."

"And how long will that take?" Trump asked.

"A year to a year and a half," Fauci answered.

Confused, Trump said that the head of Regeneron was planning to have its vaccine available within two months, which evoked laughter from the room. Regeneron was not making a vaccine but a monoclonal antibody that would treat the illness and potentially prevent worse symptoms. It also wouldn't be ready in two months.

"A little longer. A little longer," Schleifer of Regeneron said.

Amid this frenzy, Jennifer Haller, an operations manager at a Seattle tech company and a mother of two, started paying close attention. Her mother and stepfather lived not far from the nursing home where residents were getting sick, and Haller was worried about them.

On March 3, she was scrolling through Facebook at work when a friend's post stopped her. It started "Fellow Humans" and was asking for volunteers for a 14-month trial to test an experimental COVID-19 vaccine. The friend had linked to Kaiser's signup page as a favor for a nurse working on the vaccine. This was the phase 1 trial that Graham, now exclusively working from home, had set in motion and was helping oversee.

Haller barely blinked. She didn't care about the $1,100 stipend being offered to trial participants. The 43-year-old mom had stepped up because it was the right thing to do in an awful moment. She had always felt that as a white person living a middle-class life, she should give something back. She also was game for taking risks. At a previous job with a stun gun company, she had volunteered to be tased so she would know how it felt. It was worse than she expected.

So, she followed the link and filled out a form that asked only one question aside from her age and basic contact information: "Are you willing to attend 11 in-person study visits and have 4 phone visits over a 14-month period?" Yes, she was. She clicked *submit*.

Haller put it out of her mind until two days later, when her phone rang just as she sat down in a restaurant with a friend. She almost never answers calls from strangers. This time she did, and it was someone from Kaiser asking if she had 20 minutes to spare. She was about to ask if she could call back, but she sensed that calling back later would be too late. She excused herself from the table to answer questions about her health history and whether she could make all the appointments.

From there, things moved quickly. Haller went into Kaiser's offices a couple of days later for blood draws, a physical exam, and more questions. Graham was confident that the vaccine wouldn't make the disease worse. Still, side effects are common in vaccine trials, including pain at the injection site, tiredness, headache, muscle pain, chills, joint pain, swollen lymph nodes, nausea, vomiting, and fever. If any serious side effects emerged that weren't expected, the trial would be immediately suspended—even halted for good.

Haller knew she was essentially a guinea pig. It didn't faze her.

On the morning she drove to Kaiser, Haller layered a blue denim shirt over a gray tank top, knowing she would later need to bare one shoulder. Her husband, always on her case about skipping breakfast, scrambled some eggs. The night before, she had read news online that the first person would get an experimental vaccine against the novel coronavirus at Kaiser Permanente in the morning. She wondered if that person might be her.

When she arrived at the research center, not far from Seattle's iconic Space Needle, the whole world was standing by. It had been 75 days since news broke of unexplained cases of pneumonia seeming to emanate from a seafood market in Wuhan, China, and 66 days since scientists in the United States stared at the virus's genetic code and vowed to conjure a vaccine to shut it down at a record-shattering pace.

As she clicked through the news on her phone and surveyed the medical staff and journalists surrounding her that morning, she was just figuring out that she was, in fact, making history. She was the first COVID-19 vaccine trial subject.

A pharmacist in an N95 mask, protective goggles, and blue latex gloves held up a syringe labeled with an expiration a few hours away. The photographers focused their lenses on Haller's shoulder. The needle slid in.

The experiment had begun.

Shortly before Haller walked out, notifications on her cellphone started going off. The first text came from her mother. "Just received a call from Deborah Horne at KIRO TV . . . They would like to talk with you and want you to call."

Hours later she was live on CNN and MSNBC.

"I'm excited to be the first person," Haller told MSNBC's Ari Melber. "This is crazy."

"Crazy is one word for it," Melber joked, making Haller laugh.

Haller's shot marked an enormous milestone for Graham and his team. It had been only 66 days since the Chinese posted the genetic sequence. They had proven that vaccines can be ready for human tests quickly. Graham remained fully confident in the vaccine, later confiding to a reporter that at that point he already expected it to be about 80% effective. If he turned out to be right, expectations about the speed of vaccine development would change forever.

The next day, Pfizer announced that it was partnering with the mRNA competitor BioNTech to make its own COVID-19 vaccine.

In Atlanta, another snag cropped up. A courier service was trying to deliver vaccines for the Emory leg of the trial from a repository at Fisher BioServices in Germantown, Maryland. However, flights from the nation's capital kept getting canceled as the country was finally shutting down. That was a problem, because the vaccine solution would degrade if it was not kept frigidly cold.

After two days of delays at Reagan National Airport near Washington, DC, a courier service that specializes in shipping medical products gave up on flights and used its own delivery vehicles to drive down to Atlanta instead. It packed the vaccine containers in dry ice, dropping them off late the evening of March 26.

The first shot was given there the next morning—nearly two weeks after Haller had kicked off the phase 1 trial. The effort was led by Evan Anderson at Emory University. His colleague Nadine Rouphael, who was also a principal investigator on the trial, decided to be less visible because she was Lebanese and worried about the anti-immigrant sentiment within the Trump administration. Emory was able to quickly find volunteers, in part because they recruited trustworthy volunteers from past clinical trials. For example, Sean Doyle, a student at the medical school, had been in an Ebola vaccine trial. Anderson knew he could count on Doyle to follow the instructions and make all his appointments.

Anderson worked from early morning to late at night managing the trial. A key part of the analysis, being done largely at Vanderbilt University, was to test the subjects' blood samples for antibodies that would block the virus. That data was discussed in regular conference calls with Graham and the team at NIH, along with others involved in analyzing the data. The highly confidential data was kept under lock and key that Anderson could only access on his work computer.

There were moments of anxiety. In Seattle, one 29-year-old volunteer, Ian Haydon, started feeling ill 12 hours after the second dose. He had been given the highest dose of vaccine: 250 milligrams. Haydon developed a fever of 103 degrees that night and decided to go to an urgent care center. After returning home, he fainted. Two other volunteers getting the highest dose had grade 3 adverse events, meaning serious but non-life-threatening reactions. One other at a lower dose had a rash at the injection site.[7] It's not unusual for vaccines to make you feel slightly ill. But serious side effects could seriously impact people's willingness to get vaccinated. The good news was that the side effects at lower doses were not so bad.

Accounts vary on exactly when the researchers knew the trial was a success. Anderson remembers realizing that the vaccine was working by looking at the data on his computer. He told his colleague, "This is amazing. It's always exciting to see data, but especially when you're in the setting of a pandemic and it feels like the world is watching."

Jackson remembers learning that the trial was going well in a conference call.

Graham pegs it to an email from researchers at Vanderbilt on May 9. They had the results of antibodies examined from the first eight volunteers, including Haller. The tall S curves on the chart told the story—the higher the top of the S, the better. When Vanderbilt researchers took antibodies from the volunteers' blood and tested them on infected cells in the laboratory, the virus stopped replicating. Graham was expecting the vaccine to produce neutralizing antibodies, but not this strong. The highest dose tested was dropped because it produced the most side effects, including fever, but otherwise there were no major safety concerns.

In a conference call with all the key players, the team gave each other verbal high fives—"Wow!" and "Thank you!"—for a job well done.

On May 18, Moderna made the unusual decision to release those early findings in a press release. The company's stock soared that day, up 250% since December. Haller figured out she probably was one of the eight volunteers studied and was pleased that the vaccine seemed to work.

Within days, Moderna announced plans for a 600-volunteer phase 2 trial to establish a dose. That trial was barely under way when the company started crafting the phase 3 study, which would balloon to 30,000 participants, using the most promising dose from phase 1 instead of waiting for phase 2 results. Time was of the essence.

They believed the vaccine was safe; now they had to see whether it worked.

"It's the first phase 3 of a COVID vaccine in the US, it's the first phase 3 of an mRNA vaccine ever, and it's the company's first phase 3 as well," Moderna CEO Bancel told CNBC's *Squawk Box* on July 27. "So a big day came from a lot of work of a big team."

Graham wrote a commentary published on May 29, 2020, in the journal *Science* to address concerns about enhanced disease syndrome. He said it was critical to go fast with vaccine development. "Because the human population is naïve to SARS-CoV-2, the consequences of repeated epidemics will be unacceptably high mortality, severe economic disruption, and major adjustments to our way of life. Therefore, the benefit of developing an effective vaccine is very high, and even greater if it can be deployed in time to prevent repeated or continuous epidemics." But he said because the vaccine is given to otherwise healthy people, it must be safe. Antibody disease enhancement has been seen in a type of coronavirus that infects cats who had been immunized with a vaccine made from the dead virus. Graham described what happened to the children given the RSV vaccine as "vaccine-associated enhanced respiratory disease," which is distinct from antibody-dependent enhancement. Still, the children created the wrong antibodies when they were infected with the live virus that made the disease worse. The key, he said, was to assure that the vaccines produce high-quality neutralizing antibodies.[8] The FDA had already established parameters for the RSV vaccines currently under development, partly based on Graham's discoveries. Graham's suggestion was to apply those same guidelines to the COVID-19 vaccines. In early 2020, the FDA called Graham in several times to discuss how to make sure the COVID-19 vaccines would not make things worse. The FDA was satisfied with his answers.

Having fully taken the reins from Graham and his colleagues, Moderna now needed to recruit 30,000 people—a massive number set by the FDA—which Bancel estimated would take up to eight weeks. Infection rates would determine when they would have results; the worse the outbreak, the faster they would have data showing whether the vaccine worked.

"October, maybe November," Bancel said. "It's tough to know right now."

On August 6, trailing in the polls to Joe Biden, Trump suggested that the COVID-19 vaccine might be ready before Election Day. Although he had toured the NIH lab with Fauci, Graham, Corbett, and others in March, the president had previously demurred to the experts on the timing of a vaccine, focusing more on rapid tests and treatments. Outside the White House, a reporter asked if a vaccine would help him win.

"It wouldn't hurt," Trump said. "But I'm doing it not for the election; I want to save a lot of lives."

Critics jumped in, worried the vaccine was being rushed for political gain. A poll taken by CNN days later showed Americans were nervous. Only 56% said they would get the vaccine when it was approved; 40% said they wouldn't. It didn't help that the administration had called the program to accelerate development and production Operation Warp Speed.

Within a month of Trump's comments, AstraZeneca temporarily halted its trial because of a single case of spinal cord inflammation. Researchers would later determine the case was not vaccine related, but it sent shockwaves through the public.

As time went on and coronavirus deaths continued to mount, Trump started taking shots at FDA and CDC officials for playing politics by delaying the vaccine. "So we're going to have a vaccine very soon, maybe even before a very special date. You know what date I'm talking about," Trump said on September 7.

Two days later, nine vaccine makers came together to respond, pledging in a joint statement to follow "high ethical standards and sound scientific principles" and to go through the normal channels of review by the FDA. "We believe this pledge will help ensure public confidence in the rigorous scientific and regulatory process by which COVID-19 vaccines are evaluated and may ultimately be approved," the statement concluded. Looking back, Gregory Glenn, research and development president at Novavax, one of the nine companies, said, "It's really important we give gold standard evidence."

Although Graham was not involved in the public statement, he knew that the vaccine makers had no choice. Science must proceed through prescribed steps, even when lives are in the balance. The FDA could not green-light any vaccine until safety and efficacy data was available and reviewed.

Not long after Moderna began recruiting volunteers for its final trial, the head of vaccine development for Operation Warp Speed noticed a big problem: The company wasn't recruiting enough African Americans. Moncef Slaoui, a former board member at Moderna, got on the phone with Bancel and Moderna president Stephen Hoge. When Slaoui felt that Bancel wasn't listening, he began shouting. "We were shouting at each other on the phone—shouting in a respectful but very angry way, very stressed way," Slaoui said. "There was a very big tension because we need to recruit very quickly and we need to recruit the right people."

To Slaoui, it was clear that having Black and Latino participants in the trial was key to the vaccine's future success. The virus was disproportionately lethal for those communities; African Americans die of COVID-19 at nearly three times the rate of white Americans. It also was important to show that the vaccine was safe for all, a detail often neglected in clinical trials.

Moderna relented and slowed enrollment for a while. Its recruiter bought social media ads targeted to Black and Latino users. Slaoui joined a virtual town hall meeting with the Rainbow PUSH Coalition, headed by civil rights icon Jesse Jackson. "Frankly, developing a vaccine not used in the population or in a fraction of the population is the same as having no vaccine," Slaoui said at the September 24 meeting. "A vaccine on the shelf is absolutely useless."

Jackson blames the reluctance by Black and Latino people to participate on "racism in health care down through the years" in the United States, ranging from the Tuskegee experiment to poorer hospital systems in minority areas.

"The history must inform us," he said, "not paralyze us."

The pause to recruit additional participants of color lasted two weeks and resulted in a 50% increase in Black people in the trial. It also caused Moderna to lose its front-runner status to another company, Pfizer.

Graham's boss called with astonishing news on the second Sunday in November. The government's data monitoring safety board—the only group allowed to see behind the curtain and know who in the trial received a vaccine or a saltwater placebo—had completed the first analysis of 94 volunteers who had gotten COVID-19 in the Pfizer vaccine trial. To meet the FDA threshold in the first analysis with only 32 subjects, the vaccine would have to be at least 75% effective.

The data showed it had far exceeded that threshold: It was 95% effective.

As Graham hung up the phone, he took a deep breath and told his wife that the Pfizer vaccine—nearly identical to Moderna's—worked. Then he wandered over to his desk, sat down, and wept tears of relief. The type of vaccine he had been perfecting over a decade was far more successful than even he had imagined.

A documentary film crew was present when Graham called Kizzmekia Corbett the next morning at 6 a.m. to tell her the news.

GRAHAM: I am calling you because we got results yesterday and it will be announced in just a few minutes by press release.

CORBETT: *(looking nervous)* Okay. What did they say?

GRAHAM: Well, there were 95 cases total, and 90 of them were in the placebo group. Five were in the vaccines.

CORBETT: *(voice cracking)* Yeah so . . . *(breaks into laughter)* That's exciting.

GRAHAM: So, it looks like it's going to work. None of the vaccines had severe disease, which is even better news.

CORBETT: That is amazing news![9]

Graham did not yet have any word on the Moderna vaccine. But Fauci did. He had been clued in days earlier by the head of the data safety board but had promised to keep the secret until November 16. When that day came, Moderna made a nearly identical announcement: Its vaccine was 94% effective.

Corbett couldn't hide her emotions, at least on Twitter, where her profile reads "Virology. Vaccinology. Vagina-ology. Vino-ology. My tweets are my own. My science is the world's." Replying to her own tweet the day Haller was given the first vaccine in an NIH trial, Corbett had written, "I. . . . I. . . . This is . . . *tears*."

The team was recognized on the first day of December with a scientific award, the brainchild of US representative Jim Cooper, a Democrat from Nashville. Cooper sought to counter the Golden Fleece Award, which singles out silly-sounding research efforts, by pinpointing the research behind major breakthroughs. The 2020 Golden Goose Award went to Graham, Corbett, McLellan, and others for leading the effort to create a vaccine.

Corbett responded with more tweets, one aimed at colleague McLellan: "Thanks Jason! We did it!!!"

In another video post, Corbett pauses at times to hold back tears, saying: "Part of the reason why I was given essentially the wings to fly with this project is because I had a mentor who allowed me to use his resources and sit in the back of his laboratory and to ask scientific questions that a lot of other people didn't really care about."

Finally it was time to get the FDA to authorize the vaccines. Pfizer went first. Moderna followed just days later. Moderna's top scientific leaders went before a panel of 21 independent scientists on an FDA advisory panel to present their data. The December 17 meeting was held remotely via webcast—and the whole world could watch.

Their delivery was deadpan, even though this was the moment of truth. The panel's recommendation would weigh heavily on whether the FDA would agree to an emergency use authorization.

Typically, the FDA spends months analyzing data to approve a vaccine. This time, that work was done in two and a half weeks, partly because the vaccine was so extraordinarily effective. In advance, the FDA had said it was willing to give the green light to a vaccine that was 50% effective. It was even willing to stop the trial early if the vaccine proved to be 75% effective. Moderna's data, like Pfizer's before it, was even more dramatic.

Of the 196 trial participants who had COVID-19 with symptoms, only 11 had received the vaccine. All 30 cases of severe COVID-19 happened among those who got the placebo.

Later in the afternoon, an FDA doctor carefully walked the independent scientists through the side effects and adverse events. She concluded that while there were a few cases of severe side effects, such as deep muscle pain, she found no safety concerns in the data. Six people in the vaccine group had died, but none of those deaths—ranging from a heart attack to suicide—was considered vaccine related.

At the end of the day, the biggest question was whether to make it clearer that the group was voting on emergency use authorization and not regular FDA approval. In the end, no change was made to the language: "Based on the totality of scientific evidence available, do the benefits of the Moderna COVID-19 Vaccine outweigh the risks for use in individuals 18 years of age and older?"

The vote was 20 yes, with one abstention—an objection to the wording, not the vaccine. Another issue was whether to unblind the study and stop giving the placebo. The panel agreed that it would be wrong to continue the controlled study under the circumstances.

Fauci now attributes the vaccine being ready so quickly to two factors, one of them unfortunate. With the coronavirus spreading rapidly during the trial, odds dramatically increased that some of the placebo recipients in the Moderna and Pfizer trials combined were going to catch the virus quickly. That would prove that the vaccine worked in the others who received it.

The other reason, Fauci says, was the work Graham and his colleagues had been doing for years, in their own laboratories as well as in embracing mRNA and choosing it for the COVID-19 vaccine.

"I think the perspective that I had," Fauci said, "was seeing the link from the years of fundamental basic and clinical research that got us to that point, in the first week of January, when we knew that all we needed was the sequence of the new coronavirus."

Back in the lab, McLellan has found a slightly different configuration of the protein that is easier to produce and more precise. Instead of a two-proline substitution, it has six. It also appears to be more potent. Moderna and Pfizer both are reportedly exploring a next-generation vaccine that may require only one small dose, which might be ready in time to benefit poorer countries.

Although the government patented the vaccine design used by almost every vaccine maker, McLellan said as of August 2021, "We've received almost no money." Some companies had licensed the patent but were not paying any royalties yet. In keeping with its history, "Moderna is refusing to license it so they have been infringing on the patent," McLellan said. The company lost a battle in 2020 in a patent clash with Arbutus Biopharma on lipid nanoparticles used to taxi the mRNA to cells.[10] Moderna is appealing. In a statement, Moderna said, "Our improved proprietary LNP [lipid nanoparticles] formula, used to manufacture mRNA-1273 [the code name for its COVID-19 vaccine], is not covered by the Arbutus patents."[11] Now, after helping to establish Moderna as a leading vaccine maker, the government is seeking counsel from the Justice Department over whether to sue the company. "It's a lot of money," McLellan said. "All of the companies and the CEOs will be like super multibillionaires and millionaires. Most of the scientists will not get that much."

Epilogue

The end of the story could have been how we all got vaccinated and life returned to normal. But that didn't happen. Science didn't defeat the pandemic. As I'm writing this book, the Delta variant is sweeping the nation, creating a catastrophic fourth wave. The Lambda variant seems to be next in line. Hospital intensive care units are teeming with patients who chose not to get their shots. Florida hospitals were running out of oxygen.[1] Elsewhere, hospitals were running out of beds. The health care systems in at least 10 states are so full that they are in danger of collapsing.[2] Australia and New Zealand were able to tame the pandemic with masks and lockdowns.[3] But in the United States, compliance with public health measures was always lax. Vaccines were our only way out. Despite a herculean effort and a surplus of vaccine, only 57% of Americans were fully vaccinated by late October 2021. More than 93 million Americans who could get the shot were still shunning it, even as new cases and deaths soared day by day.[4] For the first time in vaccine history, science didn't win. This time, so many people have rejected the science that the virus is still winning.

The problem has only worsened as the variants have made the virus more contagious. Initially, scientists estimated that every person who got COVID-19 spread it to three others. Now, with Delta, an infected person on average makes seven other people sick. It is almost as contagious as chicken pox, a childhood disease everyone got until there was a vaccine.[5] That changes the math of herd immunity, making it much harder to achieve. Now, virtually everyone needs to be vaccinated to finally bring this pandemic to an end. "I think until 90% of the world has some immunity, we're not going to be out of the pandemic," Barney Graham said in late August 2021.

There had been so much hope early on that the art of persuasion would be enough to inoculate enough people. But now, that no longer seems plausible. President Joe Biden decided that the only way out was vaccine mandates. They are nothing new. All states require students to be vaccinated against a host of communicable diseases, such as diphtheria, tetanus, and whooping cough, to attend school. That's true even for private schools and day cares in all but a handful of states.[6] Virtually all of us have been vaccinated at some point. One of the most contagious viruses is measles. When someone is infected, they will pass it on to 90% of those in contact with them who haven't developed immunity.[7] To stop the spread of measles, 95% of us must be vaccinated. Herd immunity requires nearly everyone to cooperate. And yet, we have succeeded in preventing measles epidemics for decades without much controversy.[8]

Before the vaccine, the US averaged about 450,000 cases of measles a year.[9] Yet in 2020, there were only 13 cases.[10] This is even though there was reason not to trust the measles vaccine at one time. It was taken off the market in 1967 because it made some children more suspectable to the disease. The faulty vaccine was replaced the next year with one that worked.[11]

Employers at first seemed reluctant to require their workers to get immunized. Venerated Houston Methodist became the first hospital system in the country to insist that all its workers be inoculated. In the end, only 153 of its 26,000 workers refused and were let go.[12] A federal judge dashed the hopes of workers objecting, who argued that vaccine mandates turned them into unwilling guinea pigs and violated the Nuremberg Code for medical experimentation without consent. The judge scoffed at these notions. He said getting a vaccine authorized by the Food and Drug Administration is not the same as an experimental trial. And he added, "Equating the injection requirement to medical experimentation in concentration camps is reprehensible."[13] There wasn't much question that the laws favored vaccine mandates. There is even a 1905 Supreme Court decision upholding a vaccine mandate for all citizens of Cambridge, Massachusetts, during a smallpox epidemic.[14] Still, the Texas ruling seemed to clear the way for any employer to require its workers to get vaccinated. And the trend snowballed when the FDA finally gave full approval to the Pfizer vaccine in August 2021.

Can we achieve herd immunity without antivaxxers caving in? "I don't think so," said Anthony Fauci in an MSNBC interview. "The only way you could get to herd immunity without them is the unfortunate situation where they all wind up getting infected. And then you'll have a combination of infected individuals who have some degree of protection together with vaccinated individuals. But that's going to lead to a whole lot of suffering, a whole lot of hospitalization, and a whole lot of death . . . We can get herd immunity really easily if everyone gets vaccinated."[15]

Fauci has become the public face of the vaccines. The doctor who was criticized by his peers for steering his efforts to HIV in the 1980s went on to become one of the most respected infectious disease experts in the world. Ultimately, Fauci's obsession with an HIV vaccine paved the way for the COVID-19 vaccines. Public attitudes about him are wildly mixed. One survey in 2021 found that 68% of Americans trusted Fauci, but not the remaining 32%, many of whom vilify him with unfounded and absurd allegations of profiteering off the vaccines or even concocting the virus. Sixty-two percent of people who don't trust Fauci get their "news" from partisan sources such as Newsmax, OAN, Parler, and Gateway Pundit.[16] At a Trump campaign rally hours before Election Day, a crowd in Miami chanted "Fire Fauci." Trump played along: "Don't tell anybody but let me wait until a little bit after the election." The crowd cheered.[17]

Sadly, the public reaction to the vaccine as well as to the pandemic itself has become absurdly political. Trump continued to downplay the virus even as the death toll mounted. "Cases up because we TEST, TEST, TEST. A Fake News Media Conspiracy. Many young people who heal very fast. 99.9%. Corrupt Media conspiracy at all-time high," he tweeted just days before the election.[18] It was, of course, nonsense. Yet the idea that somehow the media and even doctors were conspiring to make the pandemic seem far worse than it was made sense to many Trump supporters. Crazy conspiracy theories ran rampant. One of the most pernicious rumors was that doctors were faking death certificates in some massive cabal to commit insurance fraud. American Medical Association president Susan Bailey took offense: "The suggestion that doctors—in the midst of a public health crisis—are overcounting COVID-19 patients or lying to line their pockets is a malicious, outrageous, and completely misguided charge."[19]

Even though Trump championed the vaccine, tried to twist arms to get it approved before the election, and even got it himself, his die-hard supporters disassociated him from it. Some even booed him when at a 2021 rally in Alabama when he said, "You know what? I believe totally in your freedoms. You got to do what you have to do, but I recommend: Take the vaccines. I did it—it's good."[20] Antivax sentiment grew so strong that some on the right looked for any alternative. Florida governor Ron DeSantis, while opposing vaccine mandates, set up more than a dozen monoclonal antibody treatment sites for anyone who was infected. The cost to the government was $1,250 per dose, as opposed to $20 for a vaccine.[21] Monoclonal antibodies may help once you get infected, but the vaccines keep you out of the hospital. The truly desperate were even buying a deworming medication for horses and livestock at farm supply stores and swallowing it, despite warnings from the FDA that it could be dangerous. "You are not a horse. You are not a cow. Seriously, y'all. Stop it," the FDA tweeted.[22]

The public response was dramatically different for the polio vaccine. When it was announced on April 12, 1955, in a University of Michigan auditorium that Jonas Salk's vaccine was as much as 90% effective against paralytic polio, Salk became and remains one of the one of the most famous scientists in history. In awarding him a presidential citation, President Dwight D. Eisenhower said, "When I think of the countless thousands of American parents and grandparents who are hereafter to be spared the agonizing fears of the annual epidemic of polio myelitis, when I think of all the agony that these people will be spared as they see loved ones suffering in bed, I must say to you I have no words in which adequately to express the thanks of myself, all the people I know, and all 164 million Americans."[23]

None of the scientists who contributed to the success of the COVID-19 vaccines have been invited to the White House or received presidential citations as of the summer of 2021. At a White House press briefing in April 2021, Fauci did offer a history of the vaccines, giving credit to Katalin Karikó, Drew Weissman, Barney Graham, Jason McLellan, Kizzmekia Corbett, and a few others.[24] Still, none of them are household names. Karikó and Weissman are often mentioned as possible future Nobel laureates. When asked about the lack of fame, Graham said, "I've never had an aspiration to be a household name. We chose the life we have. It's been very gratifying work."

Early in the pandemic, a viral video of Bill Gates seemed shockingly prescient. Wearing his signature pullover sweater, this one pink, Gates walked out onto a stage in March 2015 pushing a dolly with a 17.5-gallon barrel on it. Astonishingly, this hunter green steel drum was exactly like a one his parents—though not survivalists—had in their basement when Gates was growing up in a wealthy neighborhood of Seattle, overlooking Lake Washington. Though his parents' barrel seemed as if it might contain hazardous material, it came from the Office of Civil Defense. It was loaded with nonperishable food and other necessities in case of a nuclear holocaust. Gates was making a point. The United States had an unlimited drive and budget to fight a cold war with the Soviet Union. That war carried the fear of long-range ballistic missiles one day denotating over our heads, wiping out millions of us. We wanted to be prepared. His parents took heed.

However, Gates said there was a far more probable danger lurking. And we don't seem to want to devote even a tiny fraction of the resources of a cold war to preparing for it.

"If anything kills over 10 million people in the next few decades, it's most likely to be a highly infectious virus rather than a war. Not missiles, but microbes. Now, part of the reason for this is that we've invested a huge amount in nuclear deterrents. But we've actually invested very little in a system to stop an epidemic. We're not ready for the next epidemic."[25]

Gates would later tell Ezra Klein of Vox that the odds of them living to see a nuclear war were low while the odds of a widespread epidemic far worse than Ebola were 50 percent.[26]

If you take Gates's words literally, you could argue that SARS-CoV-2 was that widespread epidemic. You might think, then, that we could relax now. Unfortunately, that's not the case. With so many Americans refusing to get inoculated, there are dangers lurking. One is that as the virus continues to mutate, it may eventually escape the protection of existing vaccines. Whenever the virus replicates inside a human cell, it can make mistakes, producing a different version of itself. Most of these mistakes amount to nothing. But it's a game of survival of the fittest. And the mutations that manage to hold up against our antibodies the best become the variants that are most likely to thrive.

"Right now, all of these variants are appearing because the virus is trying to figure out how to infect people better," Drew Weissman said. "As more people are vaccinated, a new kind of variant will appear. And these are going to be immune evasion variants.

"Once the immune evasion variants start appearing, those are going to be really hard to get around, because the virus keeps mutating to infect people who have already been infected or vaccinated . . . Someday a variant is going to appear that the vaccines don't work against at all. And then eventually you have a brand-new pandemic."

Time is a key factor. Pfizer CEO Albert Bourla predicted there would be enough vaccine for the world to return to normal by the end of 2022. His company alone was expected to produce 3 billion doses in 2021.[27] But vaccine supply does not equal herd immunity. What's more, vaccination efforts in poorer countries were not going well. Initially, the World Health Organization hoped to have 20% of the global population vaccinated by the end of 2021. But that initiative, known as Covax, has been beset by problems. After five weeks, only 6,000 shots had been given in Chad, despite the country receiving 100,000 doses of the Pfizer vaccine. Similar problems in other countries were causing vaccine doses to expire before they could be used. And Covax was struggling to obtain all the vaccine it wanted. Even if rich countries somehow hit herd immunity, we all remain vulnerable to wily mutations until the entire globe achieves the same level of immunization.[28]

This has, believe it or not, quietly led to a new race for a vaccine. Several laboratories are now trying to develop a vaccine that would work against any variant of SARS-CoV-2. Among those already working on such a vaccine are Barney Graham, Jason McLellan, and Drew Weissman. McLellan and Graham started working on a universal vaccine in 2017. McLellan submitted a grant application to the National Institutes of Health, but it got such a poor score from anonymous peer reviewers that it got rejected. A reviewer said that the need for a coronavirus vaccine was not significant enough and that coronavirus outbreaks are contained geographically—more evidence that the peer review system is deeply flawed.

Weissman realized the need for a universal vaccine as soon as WHO declared COVID-19 a pandemic in March 2020. He teamed up with HIV researcher Barton Haynes at Duke University's school of medicine; Ralph Baric, a leading coronavirus researcher at the University of North Carolina; and others to submit a grant application to the National Institutes of Health for funding. Ironically, the peer reviewers picked by NIH also gave Weissman's grant applications a low score, dooming its chances. The reviewers said there was no need for a universal vaccine—a stunning and short-sighted opinion. Weissman said it was obvious from the start that variants would ultimately become a problem and may require a different approach. His team reapplied for the grant in 2021 and was told to expect it to be approved.

Not everyone is convinced that the current vaccines need to be replaced. As of August 2021, the vaccines remain remarkably effective at preventing hospitalization and death. A CDC study in Los Angeles County from May 1 through July 25, 2021, revealed that people who weren't vaccinated were nearly five times more likely to get infected and 29 times more likely to be hospitalized.[29] But the vaccines were less effective at preventing infection once the Delta variant swept through the country. A study of health care workers and first responders showed that the Pfizer and Moderna vaccines were 66% effective at preventing infection from the Delta variant, down from 90% effectiveness early on.[30] The Israel Ministry of Health suggested that the Pfizer vaccine was preventing infection only 39% of the time.[31] No one knows whether the problem is Delta or waning effectiveness of the vaccines. But the FDA has approved booster shots for the those age 65 and older or those with higher risks who got an mRNA vaccine. It also approved a booster for anyone who got the J&J single shot. It's possible that a booster shot could fix the problem, at least for a while.

Yet, variants remain a major concern. Researchers at the Fred Hutchinson Cancer Research Center have tried to analyze all possible variations of SARS-CoV-2. One approach is for laboratories to make one vaccine that would work against any possible variant. "You'd essentially prime the immune system so that the virus really couldn't escape from it," said Larry Corey, a virologist at the Hutch. "And it if did escape from it, it would lose its advantage to replicate. It would be what we call less fit," meaning it wouldn't spread easily.

Besides variants, the other danger lurking is an entirely new strain of coronavirus. Since 2003, the world has endured SARS-CoV-1, MERS, and SARS-CoV-2. There are thousands of coronavirus strains in bats that could one day make the leap to humans. As the population has expanded and natural habitats for animals have been destroyed, we have come in close contact with potentially new viruses that could turn into pandemics. Scientists call this spillover. And it had been happening at an alarming rate in recent decades.

"The big worry about coronavirus is that we've had three epidemics in the past 20 years," Weissman said. "So when we're hopefully done with this one, there's going to be another. And then another after that. We are far from done with coronavirus." That doesn't even take into account other viruses lurking in animals that could also create a new pandemic.

This pandemic will eventually end. But it may not end well. "We made a vaccine and the vaccine can help," Graham said. "Overall as a world, I'm not sure we're doing much better than we did in 1918." That, of course, is a reference to a flu pandemic that cost the lives of 50 million people worldwide. As of late August 2021, there 216 million confirmed cases of COVID-19 and around 4.6 million deaths worldwide. Americans account for 675,000 of those deaths—about as many Americans as died in the 1918 flu pandemic. If you're still not vaccinated, you're at high risk of catching COVID-19. You're also prolonging this pandemic. The other route to herd immunity is for billions of people to get sick. That would be a horrific outcome. Said Graham, "I hope that people generate immunity through the vaccine instead of infection, because if it's infection, there's going to be a lot more suffering and death."

WHAT ARE THE SCIENTISTS DOING NOW?

Barney Graham left the National Institutes of Health at the end of August 2021. He moved to Atlanta to be closer to family. Graham said he would take time off but at some point be involved once again in immunology. He won the Albert B. Sabin Gold Medal from the Sabin Vaccine Institute for making extraordinary contributions to vaccinology. As a government employee, he is limited to getting $150,000 a year in royalties from his patent on the two-proline substitution for spike protein. As of August 2021, however, it wasn't clear that any vaccine maker had paid any royalties.

Katalin Karikó is finally being recognized for her accomplishments. She's been profiled in several major media outlets, and she is working on her own book. She continued to work for BioNTech, where she was promoted to senior vice president in 2019. She is receiving some royalties for her modified mRNA patent that was sublicensed by Moderna and BioNTech. She has won many awards, including the Horwitz Prize for pioneering research from Columbia University.

Drew Weissman continues to work at the University of Pennsylvania. He is currently involved in a collaboration with other laboratories to make a universal coronavirus vaccine. He and Karikó make money from the patent they were never able to license themselves but was eventually sublicensed to Moderna and BioNTech. He and Karikó are now often mentioned as future Nobel Prize laureates.

Derrick Rossi left Moderna in 2014 and left Harvard in 2018. He cofounded several other biotech companies and is currently the CEO of Convelo Therapeutics, which is involved in trying to restore function to patients with neurological disorders. Rossi held on to his founder shares in Moderna, which most likely have made him a billionaire.

Luigi Warren runs his own business. He never did any more academic research after leaving Rossi's laboratory. He made millions of dollars from the patent for using mRNA to reprogram cells into IPS cells. He is very active on Twitter, where he ironically raises questions about the effectiveness of the vaccines and criticizes Anthony Fauci.

Jason McLellan has his own laboratory at the University of Texas at Austin, where he is a professor. He has created a new, more potent version of the spike protein that uses six proline substitutions as opposed to two. His laboratory is currently working on a universal coronavirus vaccine. McLellan stands to make millions of dollars from his two-proline substitution patent, but as of September 2021, no vaccine maker was paying royalties on that patent.

Kizzmekia Corbett left the National Institutes of Health in May 2021 after seven years to join the faculty of Harvard's T.H. Chan School of Public Health. She is an assistant professor and is in charge of a new Coronaviruses & Other Relevant Emerging Infectious Diseases (CoreID) Lab. She stands to make $150,000 a year from royalties on the two-proline substitution patent filed with Barney Graham and Jason McLellan in 2017 if vaccine makers ever pay for the patent.

AUTHOR'S NOTES

This book started as an effort to write a behind-the-scenes look for *USA Today* at Moderna's effort to create the first COVID-19 vaccine. I wanted to know what it would take to make a vaccine from scratch in just a year or so. No other vaccine had ever been produced that quickly. What I soon realized was that while Moderna would produce the vaccine, the science behind it had been developed by other scientists over a two-decade span.

One of my first interviews in the summer of 2020 was with Barney Graham, the deputy director of NIH's Vaccine Research Center. Graham explained that the vaccine relied on 10 years of work, first with an RSV vaccine and then with a vaccine for MERS. Moderna had collaborated with Graham on the MERS vaccine, but it was Graham and his associates who decided what would go into the vaccine.

That's when I shifted my focus to the scientists who made the vaccine possible.

The book relies heavily on many hours of interviews with the central characters, including Barney Graham, Jason McLellan, Kizzmekia Corbett, Derrick Rossi, Luigi Warren, Katalin Karikó, and Drew Weissman. Other key interviews included Anthony Fauci, John Mascola, Stephen Chanock, Jason Schrum, Robert Langer, Timothy Springer, Lisa Jackson, Paul Duprex,

George Daley, Sean Doyle, Michael Osterholm, Arthur Caplan, Larry Corey, Susan Ellenberger, Evan Anderson, Jennifer Haller, Kenechi Ejebi, Suhaib Siddiqi, Justin Quinn, Han Lee, Doug Melton, Chad Cowan, David Langer, Ryan Dietz, Peter Wright, Noubar Afeyan, Matthew Memoli, Barton Haynes, Jeremy Kim, and William Schaffner. I communicated with a few sources who did not want to be identified.

Despite numerous requests, Stéphane Bancel, the CEO of Moderna, chose not to be interviewed. Nor did any of the other executives at Moderna except for Moderna chairman Noubar Afeyan.

Most of the book is based on direct accounts by the central characters. When I relied on secondary sources, I have added footnotes. When there is no note, the information came from an original interview. The only exceptions are facts that are common knowledge.

For Bancel's life story, I relied heavily on his own words from several interviews or presentations he has given.

Whenever possible, I used documents to add detail and verify accounts. Several of my sources went back to find old emails and documents that either confirmed or refreshed their memories. In some cases, it took interviews with several sources to piece together a narrative that no one person knew in complete detail.

ACKNOWLEDGMENTS

This book would not have been possible without the contributions of so many people. In the summer of 2020, I was working on an article for *USA Today* about the race for a vaccine. As I was describing my story to one of my editors, Kelley French, she stopped me to tell me I had a book. She immediately contacted her agent, Jane Dystel, who wasted no time agreeing. Having never written a book before, their encouragement was just what I needed.

Although I interviewed a lot of people, there were a few who patiently offered countless hours to answer questions and to delve deeply back into history. I am especially grateful to Barney Graham, Jason McLellan, Derrick Rossi, Luigi Warren, Drew Weissman, and Jason Shrum for the time and attention they gave me. I would also thank Stephen Chanock, John Mascola, and Anthony Fauci, all at the National Institutes of Health, for taking the time to talk to me.

There were countless people at *USA Today* who were instrumental to its first telling, including Ramon Padilla, Jennifer Borresen, Javier Zarracina, Mitchell Thorson, Chris Powers, Mara Corbett, Karen Weintraub, and Gus Garcia-Roberts.

I turned to friends and former colleagues for advice and insights into writing a book. Most notably, I want to thank James Neff and Ken Armstrong for selflessly offering counsel.

Of course, there were friends who were aware that I was writing a book and who graciously offered their support. You know who you are, but I would especially thank Kory Kuriel and Jee Kuriel for their love and encouragement.

I want to thank Sean McGowan at Center Street, who believed in this book, as well as Alex Pappas, who took over as my editor.

Finally, I thank by wife, Kyung Song, and my amazing daughter, Talia Heath. They not only encouraged me through the difficult months of working on this book while holding down a full-time job, but they also rolled up their sleeves and helped me with research, details, and writing. Their loving encouragement was truly invaluable.

ENDNOTES

INTRODUCTION

1. "Merck & Co. (Themis) and Institut Pasteur—V591," Genetic Engineering & Biotechnology News, May 18, 2020, https://www.genengnews.com/covid-19-candidates/merck-and-co-themis-institut-pasteur-and-university-of-pittsburgh.
2. Elizabeth Weise and David Heath, "'Our Moon Shot': The World Needs a Coronavirus Vaccine ASAP," *USA Today,* April 23, 2020, https://www.usatoday.com/story/news/2020/04/23/coronavirus-vaccine-makers-covid-19-crisis/2983177001.
3. "Meet the First Person in the US to Get COVID Vaccine," KHN Morning Briefing, Kaiser Health News, last modified December 15, 2020, https://khn.org/morning-breakout/meet-the-first-person-in-the-us-to-get-covid-vaccine.
4. Jesper Pallesen et al., "Immunogenicity and Structures of a Rationally Designed Prefusion MERS-CoV Spike Antigen," *Proceedings of the National Academy of Sciences* 114, no. 35 (August 14, 2017), https://doi.org/10.1073/pnas.1707304114.
5. Barney Graham, "Prefusion Coronavirus Spike Proteins and Their Use," United States patent 20200061185, filed October 25, 2016, and issued February 27, 2020.
6. "Experts Discuss a Vaccine for COVID-19," University of Pittsburgh, October 16, 2020, https://www.pittwire.pitt.edu/news/experts-discuss-vaccine-covid-19.
7. "Middle East Respiratory Syndrome Coronavirus (MERS-CoV)," World Health Organization, accessed July 16, 2021, https://www.who.int/health-topics/middle-east-respiratory-syndrome-coronavirus-mers.

CHAPTER 1: READY FOR A PANDEMIC

1. Dan Hu et al., "Genomic Characterization and Infectivity of a Novel SARS-Like Coronavirus in Chinese Bats," *Emerging Microbes & Infections* 7, no. 1 (2018): 1–10. https://doi.org/10.1038/s41426-018-0155-5.

2. Xiao Xiao et al., "Animal sales from Wuhan wet markets immediately prior to the COVID-19 pandemic," *Scientific Reports* 11, no. 11898 (2021), https://doi.org/10.1038/s41598-021-91470-2.
3. Melissa Healy, "Government Shuts Down HIV/AIDS Vaccine Trial," *Los Angeles Times*, April 25, 2013, https://www.latimes.com/science/la-xpm-2013-apr-25-la-sci-aids-vaccine-trial-20130425-story.html.
4. "MERS Situation Update: June 2021," World Health Organization, https://applications.emro.who.int/docs/WHOEMCSR435E-eng.pdf.
5. Christina Spiropoulou, "Nipah Virus Outbreaks: Still Small but Extremely Lethal," *Journal of Infectious Diseases* 219, no. 12 (2019): 1855–1857, https://doi.org/10.1093/infdis/jiy611.
6. "China Investigates Respiratory Illness Outbreak Sickening 27," *Associated Press*, December 31, 2019, https://apnews.com/article/00c78d1974410d96fe031f67edbd86ec.
7. Robert Roos, "Estimates of SARS Death Rates Revised Upward," Center for Infectious Disease Research and Policy, May 7, 2003, https://www.cidrap.umn.edu/news-perspective/2003/05/estimates-sars-death-rates-revised-upward.
8. Amy Qin and Javier Hernández, "China Reports First Death from New Virus," *New York Times,* January 10, 2020, https://www.nytimes.com/2020/01/10/world/asia/china-virus-wuhan-death.html.
9. Sui-Lee Wee and Vivian Wang, "China Grapples with Mystery Pneumonia-Like Illness," *New York Times,* January 6, 2020, https://www.nytimes.com/2020/01/06/world/asia/china-SARS-pneumonialike.html.
10. Catherine Elton, "The Untold Story of Moderna's Race for a COVID-19 Vaccine," *Boston Magazine*, June 4, 2020, https://www.bostonmagazine.com/health/2020/06/04/moderna-coronavirus-vaccine.
11. Qin and Hernández, "China Reports First Death."

CHAPTER 2: HISTORY OF VACCINES

1. "Remembering Ali Maalin," Polio Eradication, September 26, 2018, https://polioeradication.org/news-post/remembering-ali-maalin.
2. "Smallpox," United States Food and Drug Administration, last modified March 23, 2018, https://www.fda.gov/vaccines-blood-biologics/vaccines/smallpox.
3. National Center for Health Statistics, *Health, United States 2019*, https://www.cdc.gov/nchs/data/hus/2019/004-508.pdf.
4. Rino Rappuoli et al., "Vaccines, New Opportunities for a New Society," *Proceedings of the National Academy of Sciences of the United States of America* 111, no. 34 (2014): 12288–12293, https://doi.org/10.1073/pnas.1402981111.

5. National Center for Health Statistics, *Health, United States 2019*.

6. "History of Smallpox," Centers for Disease Control and Prevention, last modified February 20, 2021, https://www.cdc.gov/smallpox/history/history.html.

7. Maya Prabhu, "Arriving at the First Vaccine: An Abridged History of Vaccination, Part 1," Gavi, April 1, 2021, https://www.gavi.org/vaccineswork/arriving-first-vaccine-abridged-history-vaccination-part-1.

8. Edward Belongia and Allison Naleway, "Smallpox Vaccine: The Good, the Bad, and the Ugly," *Clinical Medicine & Research* 1, no. 2 (2003): 87–92, https://doi.org/10.3121/cmr.1.2.87.

9. Mark Woolhouse et al., "Human Viruses: Discovery and Emergence," *Philosophical Transactions of the Royal Society of London. Series B, Biological Sciences* 367, no. 1604 (2012): 2864–2871, https://doi.org/10.1098/rstb.2011.0354.

10. "History of Smallpox," Centers for Disease Control and Prevention.

11. Stefan Riedel, "Edward Jenner and the History of Smallpox and Vaccination," *Proceedings (Baylor University. Medical Center)* 18, no. 1 (2005): 21–25, https://doi.org/10.1080/08998280.2005.11928028.

12. Mary Thompson, "Smallpox," Mount Vernon, accessed August 20, 2021, https://www.mountvernon.org/library/digitalhistory/digital-encyclopedia/article/smallpox.

13. Ciara O'Rourke, "Fact-Check: Did George Washington Order the Continental Army to Vaccinate against Smallpox?" *Austin American-Statesman*, August 2, 2021, https://www.statesman.com/story/news/politics/politifact/2021/08/02/did-george-washington-mandate-vaccines-smallpox-continental-army-during-revolutionary-war/5456106001.

14. Prabhu, "Arriving at the First Vaccine,"

15. Riedel, "Edward Jenner and the History of Smallpox and Vaccination."

16. Belongia and Naleway, "Smallpox Vaccine."

17. Riedel, "Edward Jenner and the History of Smallpox and Vaccination."

18. "'Vaccine': The Word's History Ain't Pretty," Merriam-Webster, accessed August 20, 2021, https://www.merriam-webster.com/words-at-play/vaccine-the-words-history-aint-pretty.

19. Riedel, "Edward Jenner and the History of Smallpox and Vaccination."

20. Judy Wright Lott, "Smallpox Update," Medscape, accessed August 20, 2021, https://www.medscape.com/viewarticle/472400_3.

21. Paul A. Offit, *Vaccinated: One Man's Quest to Defeat the World's Deadliest Diseases* (New York: HarperCollins, 2007), xi–xii.

22. Caroline Barranco, "The First Live Attenuated Vaccines," *Nature*, September 28, 2020, https://www.nature.com/articles/d42859-020-00008-5.

23. Barranco, "The First Live Attenuated Vaccines."

24. J. M. S. Pearce, "Louis Pasteur and Rabies: A Brief Note," *Journal of Neurology, Neurosurgery, and Psychiatry* 73, no. 82 (2002), http://dx.doi.org/10.1136/jnnp.73.1.82.

25. Maya Prabhu, "The Age of Modern Vaccines: An Abridged History of Vaccines, Part 2," Gavi, April 7, 2021, https://www.gavi.org/vaccineswork/age-modern-vaccines-abridged-history-vaccines-part-2.

26. Pearce, "Louis Pasteur and Rabies."

27. Pearce, "Louis Pasteur and Rabies."

28. Barranco, "The First Live Attenuated Vaccines."

29. "Louis Pasteur and the Development of the Attenuated Vaccine," VBI Vaccines, November 23, 2016, https://www.vbivaccines.com/evlp-platform/louis-pasteur-attenuated-vaccine.

30. "Achievements in Public Health, 1900–1999 Impact of Vaccines Universally Recommended for Children—United States, 1990–1998," Centers for Disease Control and Prevention, April 2, 1999, https://www.cdc.gov/mmwr/preview/mmwrhtml/00056803.htm.

31. Paul A. Offit, *Vaccinated*, xi–xii.

32. Nicholas Bakalar, "The Unfolding of Polio," *New York Times*, March 14, 2016, https://www.nytimes.com/2016/03/15/science/the-unfolding-of-polio.html.

33. "History of Polio," History of Vaccines, accessed August 20, 2021, https://www.historyofvaccines.org/timeline/polio.

34. Anda Baicus, "History of Polio Vaccination," *World Journal of Virology* 1, no. 4 (2012): 108–114, https://doi.org/10.5501/wjv.v1.i4.108.

35. "What Is Polio?" Centers for Disease Control and Prevention, last modified October 24, 2019, https://www.cdc.gov/polio/what-is-polio/index.htm.

36. Paul A. Offit, *The Cutter Incident: How America's First Polio Vaccine Led to the Growing Vaccine Crisis* (New Haven, CT: Yale University Press, 2007), 15–18.

37. J. K. Ferguson, "The Story of Poliomyelitis Vaccines," *Canadian Journal of Public Health* 55, no. 5 (1964): 183, http://www.jstor.org/stable/41984710.

38. "What Is the History of Polio Vaccine Use in America?," National Vaccine Information Center, accessed August 20, 2021, https://www.nvic.org/vaccines-and-diseases/polio-sv40/vaccine-history.aspx.

39. Offit, *Vaccinated*, 40–42.

40. Prabhu, "The Age of Modern Vaccines."

41. Offit, *The Cutter Incident*, 31.

42. Jerome Groopman, "'Splendid Solution' and 'Polio': March of Dimes," *New York Times*, April 10, 2005, https://www.nytimes.com/2005/04/10/books/review/splendid-solution-and-polio-march-of-dimes.html.

43. Stanley Aronson, "Jonas Salk, Medical Sleuth and Scientist," *Rhode Island Medical Journal* (2020): 87–88, http://www.rimed.org/rimedicaljournal/2020/08/2020-08-87-heritage.pdf.

44. Offit, *The Cutter Incident*, 28.

45. Offit, *Vaccinated*, 24.

46. Jennie Rothenburg Gritz, "The Anti-Vaccine Movement Is Forgetting the Polio Epidemic," *The Atlantic*, October 28, 2014, https://www.theatlantic.com/health/archive/2014/10/the-anti-vaccine-movement-is-forgetting-the-polio-epidemic/381986.

47. Mary Korr, "Q&A with Jonathan Salk, MD," *Rhode Island Medical Journal* (2020): 89–90, http://www.rimed.org/rimedicaljournal/2020/08/2020-08-89-spotlight.pdf.

48. David M. Oshinsky, *Polio: An American Story* (New York: Oxford University Press, 2005), 202.

49. "1955 Polio Vaccine Trial Announcement," University of Michigan School of Public Health, accessed August 20, 2021, https://sph.umich.edu/polio.

50. Harold Schmeck, "Dr. Jonas Salk, Whose Vaccine Turned Tide on Polio, Dies at 80," *New York Times*, June 24, 1995, https://www.nytimes.com/1995/06/24/obituaries/dr-jonas-salk-whose-vaccine-turned-tide-on-polio-dies-at-80.html.

51. Claire Gaudiani, *The Greater Good: How Philanthropy Drives the American Economy and Can Save Capitalism* (New York: Macmillan, 2004), 117.

52. Anda Baicus, "History of Polio Vaccination," *World Journal of Virology* 1, no. 4 (2012): 108–114, https://doi.org/10.5501/wjv.v1.i4.108.

53. Offit, *The Cutter Incident*.

54. Baicus, "History of Polio Vaccination."

55. "Jonas Salk," Salk, accessed August 20, 2021, https://www.salk.edu/about/history-of-salk/jonas-salk.

56. Margalit Fox, "Hilary Koprowski, Who Developed First Live-Virus Polio Vaccine, Dies at 96," *New York Times*, April 20, 2013, https://www.nytimes.

com/2013/04/21/us/hilary-koprowski-developed-live-virus-polio-vaccine-dies-at-96.html.

57. "Two Vaccines," Smithsonian National Museum of American History, accessed August 20, 2021, https://amhistory.si.edu/polio/virusvaccine/vacraces2.htm.

58. Arthur Allen, *Vaccine: The Controversial Story of Medicine's Greatest Lifesaver* (New York: W. W. Norton & Company, 2007), 187.

59. George Johnson, "Once Again, a Man with a Mission," *New York Times Magazine*, November 25, 1990, https://www.nytimes.com/1990/11/25/magazine/once-again-a-man-with-a-mission.html.

60. Paul Raeburn, "Sabin and Salk Were Linked by Achievements—and a Bitter Feud with AM-Obit-Sabin, Bjt," *Associated Press*, March 3, 1993, https://apnews.com/article/a8962960978d94bfc0d8c637a6a3fb4d.

61. Fox, "Hilary Koprowski."

62. "Two Vaccines," Smithsonian National Museum of American History.

63. Denise Grady, "As Polio Fades, Dr. Salk's Vaccine Re-emerges," *New York Times*, December 14, 1999, https://www.nytimes.com/1999/12/14/science/as-polio-fades-dr-salk-s-vaccine-re-emerges.html.

64. Samantha Vanderslott, Bernadeta Dadonaite, and Max Roser, "Vaccination," Our World in Data, last modified December 2019, https://ourworldindata.org/vaccination#vaccine-innovation.

65. Helen Mao and Shoubai Chao, "Advances in Vaccines," *Advances in Biochemical Engineering/Biotechnology* 171 (2020): 155–188, https://doi.org/10.1007/10_2019_107.

66. "Malaria," World Health Organization, April 1, 2021, https://www.who.int/news-room/fact-sheets/detail/malaria.

67. Vanderslott, Dadonaite, and Roser, "Vaccination."

68. Vanderslott, Dadonaite, and Roser, "Vaccination."

69. Dave Roos, "How a New Vaccine Was Developed in Record Time in the 1960s," History.com, accessed September 1, 2021, https://www.history.com/news/mumps-vaccine-world-war-ii.

70. Anne Harding, "Research Shows Why 1960s RSV Shot Sickened Children," Reuters, December 23, 2008, https://www.reuters.com/article/us-rsv-shot/research-shows-why-1960s-rsv-shot-sickened-children-idUSTRE4BM4SH20081223.

71. Patricio Acosta et al., "Brief History and Characterization of Enhanced Respiratory Syncytial Virus Disease," *Clinical and Vaccine Immunology* 23, no. 3 (2016), https://doi.org/10.1128/CVI.00609-15.

72. "Vaccine Types," National Institute of Allergy and Infectious Diseases, accessed September 1, 2021, https://www.niaid.nih.gov/research/vaccine-types.

CHAPTER 3: THE UNDERRATED SCIENTIST

1. P. A. Krieg and D. A. Melton, "Functional Messenger RNAs Are Produced by SP6 in Vitro Transcription of Cloned cDNAs," *Nucleic Acids Research* 12, no. 18 (1984): 7057–7070, https://www.ncbi.nlm.nih.gov/pmc/articles/PMC320142/pdf/nar00336-0141.pdf.

2. Carl Zimmer, "How Many Cells Are in Your Body?" *National Geographic*, October 23, 2013, https://www.nationalgeographic.com/science/article/how-many-cells-are-in-your-body.

3. "Sickle Cell Anemia," Mayo Clinic, last modified July 17, 2021, https://www.mayoclinic.org/diseases-conditions/sickle-cell-anemia/symptoms-causes/syc-20355876.

4. Damian Garde and Jonathan Saltzman, "The Story of mRNA: How a Once-Dismissed Idea Became a Leading Technology in the Covid Vaccine Race," *STAT*, November 10, 2020, https://www.statnews.com/2020/11/10/the-story-of-mrna-how-a-once-dismissed-idea-became-a-leading-technology-in-the-covid-vaccine-race.

5. "Soviets Put a Brutal End to Hungarian Revolution," History, last modified November 3, 2020, https://www.history.com/this-day-in-history/soviets-put-brutal-end-to-hungarian-revolution.

6. "Katalin Karikó, the Mother of the Covid-19 Vaccine, Affirms 'In Summer We Will Probably Be Able to Return to Normal Life,'" *Entrepreneur*, December 29, 2020, https://www.entrepreneur.com/article/362510.

7. Henry Kamm, "In Hungary, the Political Changes Are Tempered by Economic Fears," *New York Times*, May 15, 1989, https://www.nytimes.com/1989/05/15/world/in-hungary-the-political-changes-are-tempered-by-economic-fears.html.

8. Torontáli Zoltán, "Ha Magyarországon maradok, panaszkodó, középszerű kutató lettem volna," G7, March 31, 2020, https://g7.hu/elet/20200331/ha-magyarorszagon-maradok-panaszkodo-kozepszeru-kutato-lettem-volna.

9. Gina Kolata, "Kati Kariko Helped Shield the World from the Coronavirus," *New York Times*, April 8, 2021, https://www.nytimes.com/2021/04/08/health/coronavirus-mrna-kariko.html.

10. Susan Francia, "Olympian Susan Francia on How Her Mother Helped Develop the COVID-19 Vaccines and Their American Dream," ESPN, June 28, 2021, https://www.espn.com/olympics/story/_/id/31707795/

olympian-susan-francia-how-mother-helped-develop-covid-19-vaccines-their-american-dream.

11. Retro Report, "The Race to Sequence the Human Genome and What It Means," April 3, 2018, https://www.youtube.com/watch?v=gclpzqdV7hss.
12. Gina Kolata, "Kati Kariko Helped Shield the World."
13. Gina Kolata, "Kati Kariko Helped Shield the World."
14. Gina Kolata, "Kati Kariko Helped Shield the World."
15. David Cox, "How mRNA Went from a Scientific Backwater to a Pandemic Crusher," *Wired*, December 2, 2020, https://www.wired.co.uk/article/mrna-coronavirus-vaccine-pfizer-biontech.
16. David Cox, "How mRNA Went from a Scientific Backwater."
17. John Boockvar, "Neurosurgery Department History," Penn Medicine, accessed August 3, 2021, https://www.pennmedicine.org/departments-and-centers/neurosurgery/about-us/history.
18. Jordan Pober and William Sessa, "Evolving Functions of Endothelial Cells in Inflammation," *Nature Reviews Immunology* 7, (2007): 803–815, https://doi.org/10.1038/nri2171.
19. "False: Dr. Robert Malone Invented mRNA Vaccines," Logically, July 18, 2021, https://www.logically.ai/factchecks/library/3aa2eefd.

CHAPTER 4: THE COLLABORATION

1. Anthony Fauci, "Dr. Fauci on 30 years of AIDS," HIV Gov, August 30, 2011, video, @1:29, https://www.youtube.com/watch?v=gzBJ-y-B_6I&t=89s.
2. "About the Morbidity and Mortality Weekly Report (MMWR) Series," CDC, last modified July 2, 2021, https://www.cdc.gov/mmwr/about.html.
3. "Pneumocystis Pneumonia—Los Angeles," CDC, June 5, 1981, https://www.cdc.gov/mmwr/preview/mmwrhtml/june_5.htm.
4. Anthony Fauci, "Q&A," C-SPAN, January 8, 2015, video, https://www.c-span.org/video/?323680-1/qa-dr-anthony-fauci.
5. "Kaposi's Sarcoma and Pneumocystis Pneumonia among Homosexual Men—New York City and California," *Morbidity and Mortality Weekly Report* 30, no. 25 (1981): 305–316, https://stacks.cdc.gov/view/cdc/1265.
6. Anthony Fauci, "Q&A."
7. Anthony Fauci, "Dr. Fauci on 30 Years of AIDS," @3:00.
8. Michael Specter, "How Anthony Fauci Became America's Doctor," *New Yorker*, April 20, 2020, https://www.newyorker.com/magazine/2020/04/20/how-anthony-fauci-became-americas-doctor.

9. Anthony Fauci, "Q&A."

10. Anthony Fauci, "Dr. Fauci on 30 Years of AIDS," @7:18.

11. Anthony Fauci, "The Syndrome of Kaposi's Sarcoma and Opportunistic Infections: An Epidemiologically Restricted Disorder of Immunoregulation," *Annals of Internal Medicine* 96, no. 6 (1982): 777–779, https://doi.org/10.7326/0003-4819-96-6-777.

12. Randy Shilts, *And the Band Played On* (New York: St. Martin's Press, 1987), 171.

13. German Lopez, "The Reagan Administration's Unbelievable Response to the HIV/AIDS Epidemic," Vox, last modified December 1, 2016, https://www.vox.com/2015/12/1/9828348/ronald-reagan-hiv-aids.

14. Randy Shilts, *And the Band Played On*, 328.

15. Jon Cohen, "At Gathering of HIV/AIDS Pioneers, Raw Memories Mix with Current Conflicts," *Science*, October 26, 2016, https://www.sciencemag.org/news/2016/10/gathering-hivaids-pioneers-raw-memories-mix-current-conflicts.

16. "Current Trends Update: Acquired Immunodeficiency Syndrome (AIDS)—United States," Centers for Disease Control and Prevention, November 30, 1984, https://www.cdc.gov/mmwr/preview/mmwrhtml/00000442.htm.

17. Anthony Fauci, "Dr. Fauci on 30 Years of AIDS," @16:48.

18. Bridget Velasquez, "Lexington Native's mRNA Research Leads to Coronavirus Vaccine," *Colonial Times Magazine*, 2021, https://colonialtimesmagazine.com/lexington-natives-mrna-research-leads-to-coronavirus-vaccine.

19. "RNA Delivery for Dendritic Cell HIV Antigen Presenation," National Institutes of Health RePORT database, https://reporter.nih.gov/search/wsE2oFYQGU2xH6ca70Rt1A/project-details/6020034.

20. Anthony Fauci, "Dr. Fauci on 30 Years of AIDS," @12:05.

21. Ananya Mandal, "What Are Dendritic Cells?" News Medical Life Sciences, last modified February 26, 2019, https://www.news-medical.net/health/What-Are-Dendritic-Cells.aspx.

22. "Ralph M. Steinman Biographical," Nobel Prize, accessed August 21, 2021, https://www.nobelprize.org/prizes/medicine/2011/steinman/biographical.

23. "The Business of Biotech," Smithsonian National Museum of American History, accessed August 21, 2021, https://americanhistory.si.edu/collections/object-groups/birth-of-biotech/the-business-of-biotech.

24. "Recombinant Proteins Market Size 2021 with CAGR of 8.1%, Top Growth Companies: Abcam PLC, R&D Systems, Miltenyi Biotec, and, End-User,

SWOT Analysis in Industry 2026," MarketWatch, June 16, 2021, https://www.marketwatch.com/press-release/recombinant-proteins-market-size-2021-with-cagr-of-81-top-growth-companies-abcam-plc-rd-systems-miltenyi-biotec-and-end-user-swot-analysis-in-industry-2026-2021-06-16.

25. "Antigen Specific Activation and HIV Replication," National Institutes of Health RePORT database, https://reporter.nih.gov/search/wsE2oFYQGU2xH6ca70Rt1A/project-details/2791699.

26. "Adjuvants and Vaccines," Centers for Disease Control and Prevention, last modified August 14, 2020, https://www.cdc.gov/vaccinesafety/concerns/adjuvants.html.

27. Drew Weissman et al., "HIV Gag mRNA Transfection of Dendritic Cells (DC) Delivers Encoded Antigen to MHC Class I and II Molecules, Causes DC Maturation, and Induces a Potent Human In Vitro Primary Immune Response," *Journal of Immunology* 168, no. 8 (2000): 4710–4717, https://doi.org/10.4049/jimmunol.165.8.4710.

28. David Brown, "Hepatitis Drug May Also Have Caused a Patient's Death in January," *Washington Post*, July 9, 1993, https://www.washingtonpost.com/archive/politics/1993/07/09/hepatitis-drug-may-also-have-caused-a-patients-death-in-january/5d4fe096-81f9-42ca-b73c-f3e28264ecda.

29. Marlene Cimons, "Fifth Patient Dies in Drug Test Gone Awry: Medicine: Woman, 37, Had Received Two Liver Transplants. The 15-Person NIH Experiment Had Been Studying a Promising Treatment for Hepatitis B," *Los Angeles Times*, September 1, 1993, https://www.latimes.com/archives/la-xpm-1993-09-01-mn-30127-story.html.

30. "LiverTox: Clinical and Research Information on Drug-Induced Liver Injury [Internet]" (Bethesda, MD: National Institute of Diabetes and Digestive and Kidney Diseases; 2012–), Nucleoside Analogues, May 1, 2020, PMID 31644243.

31. Meir Rinde, "The Death of Jesse Gelsinger, 20 Years Later," Science History Institute, June 4, 2019, https://www.sciencehistory.org/distillations/the-death-of-jesse-gelsinger-20-years-later.

32. Rinde, "The Death of Jesse Gelsinger."

33. Robin Fretwell Wislon, "The Death of Jesse Gelsinger: New Evidence of the Influence of Money and Prestige in Human Research," *Washington & Lee University School of Law Scholarly Commons,* 296, https://scholarlycommons.law.wlu.edu/cgi/viewcontent.cgi?article=1125&context=wlufac.

34. Katalin Karikó et al., "Suppression of RNA Recognition by Toll-like Receptors: The Impact of Nucleoside Modification and the Evolutionary Origin of RNA," *Cell Press* 23, no. 2 (2005): 165–175, https://doi.org/10.1016/j.immuni.2005.06.008.

35. Katalin Karikó and Drew Weissman, "RNA Containing Modified Nucleosides and Methods of Use Thereof," United States Patent US8278036B2, filed August 21, 2006, and issued October 2, 2012.

36. Connor Simpson, "Lance Armstrong Admits Using EPO and Blood Doping to Oprah Winfrey," *The Atlantic*, January 17, 2013, https://www.theatlantic.com/culture/archive/2013/01/live-lance-armstrong-admit-doping-oprah-winfrey/319335.

37. "RNARX," Small Business Innovation Research, accessed August 21, 2021, https://www.sbir.gov/node/294073.

38. Katalin Karikó et al., "Incorporation of Pseudouridine into mRNA Yields Superior Nonimmunogenic Vector with Increased Translational Capacity and Biological Stability," *Molecular Therapy* 16, no. 11 (2008): 1833–1840, https://doi.org/10.1038/mt.2008.200.

39. Katlin Karikó et al., "Incorporation of Pseudouridine into mRNA."

CHAPTER 5: SCIENTIFIC SABOTAGE

1. Alice Park, "People Who Mattered: Derrick Rossi," *Time*, http://content.time.com/time/specials/packages/article/0,28804,2036683_2036767_2037437,00.html.

2. Alice Park, "Time 100: Derrick Rossi," *Time*, April 21, 2011, http://content.time.com/time/specials/packages/article/0,28804,2066367_2066369_2066502,00.html.

3. Megan Scudellari, "How iPS Cells Changed the World," *Nature* 534, no. 7607 (2016): 310–312, https://doi.org/10.1038/534310a.

4. Nicholas Wade, "Biologists Make Skin Cells Work Like Stem Cells," *New York Times*, June 6, 2007, https://www.nytimes.com/2007/06/06/science/06cnd-cell.html.

5. "Fact Sheet on Presidential Executive Order: Removing Barriers to Responsible Scientific Research Involving Human Stem Cells," Obama White House Archives, accessed August 21, 2021, https://obamawhitehouse.archives.gov/realitycheck/the-press-office/fact-sheet-presidential-executive-order.

6. Michael McCarthy, "Genes from 'Fossil' Virus in Human DNA Found to Be Active," University of Washington School of Medicine, November 4, 2019, https://newsroom.uw.edu/news/genes-%E2%80%98fossil%E2%80%99-virus-human-dna-found-be-active#.

7. Luigi Warren et al., "Transcriptional Instability Is Not a Universal Attribute of Aging," *Aging Cell* 6, no. 6 (2007): 775–782, https://doi.org/10.1111/j.1474-9726.2007.00337.x.

8. "Anthrax Investigation: Closing a Chapter," Federal Bureau of Investigation, August 6, 2008, https://archives.fbi.gov/archives/news/stories/2008/august/amerithrax080608a.

9. Brendan Maher, "ISSCR 2008: It's 'Shinyamania,'" *Nature*, June 12, 2008, http://blogs.nature.com/inthefield/2008/06/isscr_2008_its_shinyamania_1.html.

10. "RNA Molecules Are Masters of Their Own Destinies," National Science Foundation, February 8, 2021, https://www.nsf.gov/discoveries/disc_summ.jsp?cntn_id=302074.

11. *Warren v. The Children's Hospital Corporation*, Massachusetts District Court, case 1:2017cv12472 (December 19, 2017).

12. *Warren v. The Children's Hospital Corporation*.

13. "About Us," CureVac, accessed August 21, 2021, https://www.curevac.com/en/about-us.

14. Luigi Warren et al., "Highly Efficient Reprogramming to Pluripotency and Directed Differentiation of Human Cells with Synthetic Modified mRNA," *Cell Stem Cell* 7, no. 5 (2010): 618–630, https://doi.org/10.1016/j.stem.2010.08.012.

15. *Warren v. The Children's Hospital Corporation*.

16. Catherine Elton, "Does Moderna Therapeutics Have the NEXT Next Big Thing?," *Boston Magazine*, February 26, 2013, https://www.bostonmagazine.com/health/2013/02/26/moderna-therapeutics-new-medical-technology/2/.

17. "New 2010 impact factors—Cell Press Journals Deliver Top Class Performances," EurekAlert, June 29, 2011, https://www.eurekalert.org/news-releases/891243.

18. "Noubar Afeyan," Flagship Pioneering, accessed August 21, 2021, https://www.flagshippioneering.com/people/noubar-afeyan.

19. Noubar Afeyan, "Interview with Noubar Afeyan of Flagship Pioneering, Full Interview," HarvardX, December 15, 2020, video, @3:21, https://www.youtube.com/watch?v=V5TrUzA30Yo&t=201s.

20. Jeffrey Brainard and Jia You, "What a Massive Database of Retracted Papers Reveals about Science Publishing's 'Death Penalty,'" *Science*, October 25, 2018, https://www.sciencemag.org/news/2018/10/what-massive-database-retracted-papers-reveals-about-science-publishing-s-death-penalty.

21. Warren et al., "Highly Efficient Reprogramming."

22. "BMS Acquires iPierian for Up to $725M," Genetic and Biotechnology News, April 29, 2014, https://www.genengnews.com/news/bms-acquires-ipierian-for-up-to-725m.

CHAPTER 6: WHO FOUNDED MODERNA?

1. Stephane Bancel, "What If mRNA Could Be a Drug? | Stephane Bancel | TEDxBeaconStreet," TEDx Talks, December 27, 2013, video, @14:11, https://www.youtube.com/watch?v=T4-DMKNT7xI.
2. Derrick Rossi and Luigi Warren, "Compositions, Kits, and Methods for Making Induced Pluripotent Stem Cells Using Synthetic Modified RNAs," United States patent US8802438B2, filed April 15, 2011, and issued August 12, 2014.
3. Luigi Warren et al., "Highly Efficient Reprogramming to Pluripotency and Directed Differentiation of Human Cells with Synthetic Modified mRNA," *Cell Stem Cell* 7, no. 5 (2010): 618–630, https://doi.org/10.1016/j.stem.2010.08.012.
4. Derrick Rossi, "Derrick Rossi: Modified RNAs Advance Stem Cell Field," Boston Children's Hospital, September 30, 2010, video, @4:34, https://www.youtube.com/watch?v=pfYvuZdOhPs.
5. "CureVac's Ingmar Hoerr Tells Us Why He Returned to Take On a Pandemic; NGM Loses a President," *Endpoints News*, March 13, 2020, https://endpts.com/curevacs-ingmar-hoerr-tells-us-why-he-returned-to-take-on-a-pandemic-ngm-loses-a-president.
6. Ryan Cross, "Can mRNA Disrupt the Drug Industry?" *Chemical & Engineering News*, September 3, 2018, https://cen.acs.org/business/start-ups/mRNA-disrupt-drug-industry/96/i35.
7. Scott Hensley, "Researchers Take Another Step toward Stem Cells without Embryos," *NPR*, October 1, 2010, https://www.npr.org/sections/health-shots/2010/10/01/130262558/researchers-take-another-step-toward-stem-cells-without-embryos.
8. Erin Kutz, "ModeRNA, Stealth Startup Backed by Flagship, Unveils New Way to Make Stem Cells," Xconomy, October 4, 2010, https://xconomy.com/boston/2010/10/04/moderna-stealth-startup-backed-by-flagship-unveils-new-way-to-make-stem-cells.
9. Erin Kutz, "ModeRNA, Stealth Startup."
10. Alice Park, "People Who Mattered: Derrick Rossi," *Time*, http://content.time.com/time/specials/packages/article/0,28804,2036683_2036767_2037437,00.html.
11. "Flagship VentureLabs Company ModeRNA Therapeutics, and co-founder, Derrick Rossi, both Cited by Time Magazine for Significant Advances and Contributions in 2010," Flagship Ventures, December 9, 2010, https://web.archive.org/web/20120103235926/http://www.flagshipventures.com/about/news/flagship-venturelabs-company-moderna-therapeutics-and-co-founder-derrick-rossi-both-cited.

12. Moderna Inc. Form S-1 Registration Statement, Securities and Exchange Commission, November 9, 2018, https://www.sec.gov/Archives/edgar/data/0001682852/000119312518323562/d577473ds1.htm.

13. Moderna Inc. S-1 Registration Form, "Patent Sublicense Agreement," Exhibit 10.8, June 26, 2017, https://www.sec.gov/Archives/edgar/data/1682852/000119312518323562/d577473dex108.htm.

14. Moderna Inc. Form 10-K, Securities and Exchange Commission, December 31, 2018, https://investors.modernatx.com/node/7421/html.

15. Moderna Inc. Form 10-Q, Securities and Exchange Commission, Quarter ending March 31, 2021, https://www.sec.gov/Archives/edgar/data/0001682852/000168285221000017/mrna-20210331.htm.

CHAPTER 7: ENTER STÉPHANE BANCEL

1. Stéphane Bancel, "Harvard i-lab | The Other Side Speaker Series w/ Stéphane Bancel," Harvard Innovation Labs, April 19, 2016, video, @22:33, https://www.youtube.com/watch?v=-P53wVGfvjw&t=1353s.

2. "Enquête: Stéphane Bancel, le Français derrière Moderna Therapeutics qui a promis à Donald Trump un vaccin contre le coronavirus," *Vanity Fair* (France), November 16, 2020, https://www.vanityfair.fr/pouvoir/business/story/stephane-bancel-et-moderna-therapeutics-lentreprise-qui-a-promis-a-donald-trump-un-vaccin-contre-le-coronavirus/12140.

3. Damian Garde, "Ego, Ambition, and Turmoil: Inside One of Biotech's Most Secretive Startups," *Stat*, September 13, 2016, https://www.statnews.com/2016/09/13/moderna-therapeutics-biotech-mrna.

4. Damian Garde, "Ego, Ambition, and Turmoil."

5. Luke Timmerman and Stéphane Bancel, "Ep11: Stephane Bancel," *The Long Run with Luke Timmerman*, January 31, 2018, produced by HeadStepper Media, podcast, https://podcasts.apple.com/us/podcast/the-long-run-with-luke-timmerman/id1282838969?i=1000401100978.

6. Stéphane Bancel, "Harvard i-lab | The Other Side Speaker Series w/ Stéphane Bancel."

7. Luke Timmerman and Stéphane Bancel, "Ep11: Stephane Bancel."

8. Stéphane Bancel, "Harvard i-lab | The Other Side Speaker Series w/ Stéphane Bancel."

9. Marcy l'Etoile, "The Board of Directors of bioMérieux, Chaired by Alain Mérieux, Has Appointed Stéphane Bancel Directeur Général Délégué (Chief Executive Officer) of bioMérieux starting January 1, 2007," bioMérieux, December 19, 2007, https://www.biomerieux.com/en/board-directors-biomerieux-chaired-alain-merieux-has-appointed-stephane-bancel-directeur-general.

10. Luke Timmerman and Stéphane Bancel, "Ep11: Stephane Bancel."
11. Stéphane Bancel, "Transforming Medicine | Stéphane Bancel | TEDxBeaconStreet," TEDx Talks, January 4, 2017, video, https://www.youtube.com/watch?v=BwdpElhTmV4.
12. Stéphane Bancel, "Harvard i-lab | The Other Side Speaker Series w/ Stéphane Bancel."
13. Eli Lilly, "FDA Issues Warning Letter after Inspection, March 8, 2001, https://investor.lilly.com/static-files/0e268b26-fac5-4b25-899f-999fb6b84389.
14. Luke Timmerman and Stéphane Bancel, "Ep11: Stephane Bancel."
15. Stéphane Bancel, "Harvard i-lab | The Other Side Speaker Series w/ Stéphane Bancel."
16. Stéphane Bancel, "Harvard i-lab | The Other Side Speaker Series w/ Stéphane Bancel."
17. Noelle Mennella, "Interview—BioMerieux CEO Sees Growth Resurgence in US," Reuters, last modified July 23, 2008, https://cn.reuters.com/article/sppage015-l23103165-oishe/interview-biomerieux-ceo-sees-growth-resurgence-in-u-s-idUKL2310316520080723.
18. Luke Timmerman and Stéphane Bancel, "Ep11: Stephane Bancel."
19. "BG Medicine, Inc. Appoints Stephane Bancel as Executive Chairman of the Board," Intrado, July 27, 2011, https://www.globenewswire.com/news-release/2011/07/27/452233/227597/en/BG-Medicine-Inc-Appoints-Stephane-Bancel-as-Executive-Chairman-of-the-Board.html.
20. Stéphane Bancel, "In Conversation with Stephane Bancel on Jan. 20, 2021," HBS Club of the GCC, January 21, 2021, video, @8:57, https://www.youtube.com/watch?v=tCwYlVUxOUI&t=537s.
21. Stéphane Bancel, "Harvard i-lab | The Other Side Speaker Series w/ Stéphane Bancel," @31:07.
22. Stéphane Bancel, "Harvard i-lab | The Other Side Speaker Series w/ Stéphane Bancel."
23. Stéphane Bancel, "Harvard i-lab | The Other Side Speaker Series w/ Stéphane Bancel."
24. Stephane Bancel, "What If mRNA Could Be a Drug? | Stephane Bancel | TEDxBeaconStreet," TEDx Talks, December 27, 2013, video, https://www.youtube.com/watch?v=T4-DMKNT7xI.
25. Stephane Bancel, "What If mRNA Could Be a Drug?"
26. Stéphane Bancel, "Harvard i-lab | The Other Side Speaker Series w/ Stéphane Bancel."

27. Stéphane Bancel, "Harvard i-lab | The Other Side Speaker Series w/ Stéphane Bancel."
28. Stéphane Bancel, "Harvard i-lab | The Other Side Speaker Series w/ Stéphane Bancel."
29. Stéphane Bancel, "In Conversation with Stephane Bancel."
30. Moderna Inc. Form S-1 Registration Statement, Securities and Exchange Commission, August 30, 2018, https://www.sec.gov/Archives/edgar/data/1682852/000095012318009220/filename1.htm.
31. Stéphane Bancel, "Transforming Medicine | Stéphane Bancel | TEDxBeaconStreet."
32. Moderna Inc., Schedule 14A 2021, Securities and Exchange Commission, https://www.sec.gov/Archives/edgar/data/0001682852/000130817921000031/lmrn2021_def14a.htm.
33. "Enquête: Stéphane Bancel."
34. "Moderna Reviews," Glassdoor, last modified August 16, 2021, https://www.glassdoor.com/Reviews/Moderna-Reviews-E453959_P2.htm?sort.sortType=RD&sort.ascending=true&filter.iso3Language=eng.
35. "Flagship VentureLabs Company ModeRNA Therapeutics, and Co-Founder, Derrick Rossi, Both Cited by Time Magazine for Significant Advances and Contributions in 2010," Flagship Ventures, December 9, 2010, https://web.archive.org/web/20120103235926/http://www.flagshipventures.com/about/news/flagship-venturelabs-company-moderna-therapeutics-and-co-founder-derrick-rossi-both-cited.
36. "AstraZeneca and Moderna Therapeutics Announce Exclusive Agreement to Develop Pioneering Messenger RNA Therapeutics™ in Cardiometabolic Diseases and Cancer," AstraZeneca, March 21, 2013, https://www.astrazeneca.com/media-centre/press-releases/2013/astrazeneca-moderna-therapeutics-cardiometabolic-diseases-cancer-treatment-21032013.html.
37. Luke Timmerman and Stéphane Bancel, "Ep11: Stephane Bancel."
38. Stéphane Bancel, "What If mRNA Could Be a Drug?"
39. Lior Zangi et al., "Modified mRNA Directs the Fate of Heart Progenitor Cells and Induces Vascular Regeneration after Myocardial Infarction," *Nature Biotechnology* 31, no. 10 (2013): 898–907, https://doi.org/10.1038/nbt.2682.
40. Zangi et al., "Modified mRNA Directs the Fate."
41. Li-Ming Gan et al., "Intradermal Delivery of Modified mRNA Encoding VEGF-A in Patients with Type 2 Diabetes," *Nature Communications* 10, no. 871 (2019), https://doi.org/10.1038/s41467-019-08852-4.

42. "Phase 1 Data Published in Nature Communications Show Potential of mRNA Encoding VEGF-A as a Regenerative Therapeutic," Business Wire, February 20, 2019, https://www.businesswire.com/news/home/20190220005139/en/Phase-1-Data-Published-in-Nature-Communications-Show-Potential-of-mRNA-Encoding-VEGF-A-as-a-Regenerative-Therapeutic.

43. Ryan Cross, "Moderna and AstraZeneca's mRNA Therapy for Heart Regeneration Passes Phase I Safety Test," *Chemical and Engineering News,* February 20, 2019. https://cen.acs.org/business/Moderna-AstraZenecas-mRNA-therapy-heart/97/i8.

44. "DARPA Awards Moderna Therapeutics a Grant for up to $25 Million to Develop Messenger RNA Therapeutics™," Moderna, October 2, 2013, https://investors.modernatx.com/news-releases/news-release-details/darpa-awards-moderna-therapeutics-grant-25-million-develop.

45. "The Shock of Sputnik," National Park Service, accessed August 26, 2021, https://www.nps.gov/articles/mimiarmsrace-01.htm.

46. Stéphane Bancel, "Transforming Medicine | Stéphane Bancel | TEDxBeaconStreet."

CHAPTER 8: TACKLING A CHILDHOOD DISEASE

1. Jared Hopkins, "Bill and Barney, Two Old College Pals, Help Save the World from Covid-19," *Wall Street Journal*, May 25, 2021, https://www.wsj.com/articles/two-college-pals-reunite-after-50-years-in-race-for-covid-19-vaccines-11621956170.

2. Clinton Colmenares, "Clear objectives—Barney Graham Leaves Vanderbilt for NIH, but His Feet Stay Planted," Vanderbilt Health, accessed August 26, 2021, https://reporter.newsarchive.vumc.org/index.html?ID=1523.

3. "Respiratory Syncytial Virus Infection (RSV): Trends and Surveillance," Centers for Disease Control and Prevention, last modified December 18, 2020, https://www.cdc.gov/rsv/research/us-surveillance.html.

4. Robert Pear, "Grants for Medical Research to Be Cut by Administration," *New York Times*, January 21, 1985, https://www.nytimes.com/1985/01/21/us/grants-for-medical-research-to-be-cut-by-administration.html.

5. B. Lee Ligon, "Robert M. Chanock, MD: A Living Legend in the War against Viruses," *Seminars in Pediatric Infectious Diseases* 9, no. 3 (1998): 258–269, https://doi.org/10.1016/S1045-1870(98)80040-X.

CHAPTER 9: TRAGIC TRIAL

1. Lawrence Altman, "Dr. Robert M. Chanock, Prominent Virologist, Dies at 86," *New York Times*, August 4, 2010, https://www.nytimes.com/2010/08/05/health/05chanock.html.
2. Robert Chanock, interview by Peggy Dillon, National Institutes of Health, January 11, 2001, https://history.nih.gov/display/history/Chanock%2C+Robert+2001+A.
3. Robert Chanock, interview by Peggy Dillon.
4. Robert Chanock, interview by Peggy Dillon.
5. Robert Chanock, interview by Peggy Dillon.
6. Claudia Levy, "Robert H. Parrott, 76," *Washington Post*, December 28, 1999, https://www.washingtonpost.com/archive/local/1999/12/28/robert-h-parrott-76/14032e4b-ccd3-41d3-bde8-51d0b50e9f75.
7. Caroline Hall et al., "The Burden of Respiratory Syncytial Virus Infection in Young Children," *New England Journal of Medicine* 360, no. 6 (2009): 588–598, https://doi.org/10.1056/NEJMoa0804877.
8. Stephen Chanock, interview by author, August 3, 2021.
9. "Junior Village Is an—Outrage," *Washington Post*, January 20, 1971.
10. Aaron Latham, "Jr. Village: Dumping Ground," *Washington Post*, January 17, 1971.
11. Congressional Record—House, June 26, 1962, page 11727.
12. Eric Pianin, "A Senator's Shame," *The Washington Post*, June 19, 2005, https://web.archive.org/web/20071117055016/https://www.washingtonpost.com/wp-dyn/content/article/2005/06/18/AR2005061801105_pf.html.
13. Aaron Latham, "Jr. Village: Dumping Ground."
14. Aaron Latham, "Jr. Village: Dumping Ground."
15. Diane Bernard, "It Was Created as a Refuge for Needy Kids. Instead, They Were Raped and Drugged," *Washington Post*, May 18, 2019, https://www.washingtonpost.com/history/2019/05/18/it-was-created-refuge-needy-kids-instead-they-were-raped-drugged.
16. Robert Chanock, interview by Peggy Dillon.
17. Susan E. Lederer, "Orphans as Guinea Pigs," *In the Name of the Child* (London: Routledge, 1992).
18. Laura Johannes, "MIT, Quaker Oats Settle Lawsuit Involving Radioactive Experiment," *Wall Street Journal*, January 2, 1998, https://www.wsj.com/articles/SB883585397204864500.

19. David Morens and Anthony Fauci, "In Memoriam: Albert Z. Kapikian, MD, 1930–2014," *Journal of Infectious Diseases* 211, no. 8 (2015): 1199–1201, https://doi.org/10.1093/infdis/jiv034.

20. "Vaccine Developer Albert Z. Kapikian, MD, Awarded the Sabin Gold Medal," Sabin Vaccine Institute, accessed August 28, 2021, https://www.sabin.org/updates/news/vaccine-developer-albert-z-kapikian-md-awarded-sabin-gold-medal.

21. Morens and Fauci, "In Memoriam: Albert Z. Kapikian, MD."

22. Albert Kapikian et al., "An Epidemiologic Study of Altered Clinical Reactivity to Respiratory Syncytial (RS) Virus Infection in Children Previously Vaccinated with an Inactivated RS Virus," *American Journal of Epidemiology* 89, no. 4 (1969): 405–421, https://doi.org/10.1093/oxfordjournals.aje.a120954.

23. Kapikian et al., "An Epidemiologic Study of Altered Clinical Reactivity."

24. Brigitte Fauroux et al., "The Burden and Long-Term Respiratory Morbidity Associated with Respiratory Syncytial Virus Infection in Early Childhood," *Infectious Diseases and Therapy* 6, no. 2 (2017): 173–197, https://doi.org/10.1007/s40121-017-0151-4.

25. Natalie Mazur et al., "The Respiratory Syncytial Virus Vaccine Landscape: Lessons from the Graveyard and Promising Candidates," *The Lancet* 18, no. 10 (2018): 295–311, https://doi.org/10.1016/S1473-3099(18)30292-5.

26. Lynne Peeples, "News Feature: Avoiding Pitfalls in the Pursuit of a COVID-19 Vaccine," *Proceedings of the National Academy of Sciences of the United States of America* 117, no. 15 (2020): 8218–8221, https://doi.org/10.1073/pnas.2005456117.

CHAPTER 10: VACCINE RESEARCH CENTER

1. Bill Clinton, "Spring Commencement Address," May 18, 1997, Hughes Memorial Stadium, Baltimore, video, https://commencement.morgan.edu/past-ceremonies/1997-commencement.

2. David Folkenflik, "Clinton Sets 10-Year Goal to Halt AIDS President, at Morgan Commencement, Vows Discovery of Vaccine; Foresees 'Age of Biology'; Ceremony Is Attended by Governor, Mayor, Other Top Md. Leaders," *Baltimore Sun*, May 19, 1997, https://www.baltimoresun.com/news/bs-xpm-1997-05-19-1997139014-story.html.

3. Sandra Crouse Quinn and Stephen B. Thomas, "Presidential Apology for the Study at Tuskegee," Britannica, last modified August 17, 2001, https://www.britannica.com/topic/Presidential-Apology-for-the-Study-at-Tuskegee-1369625.

4. "[Commencement Address: Bill Clinton at Morgan State University] Page: 3 of 6," Portal to Texas History, accessed August 28, 2021, https://texashistory.unt.edu/ark:/67531/metadc786192/m1/3.

5. Bill Clinton, "Spring Commencement Address."

6. Christine Case-Lo, "HIV Vaccine: How Close Are We?" Healthline, last modified June 16, 2020, https://www.healthline.com/health/hiv-aids/vaccine-how-close-are-we.

7. Fernando Polack, "Atypical Measles and Enhanced Respiratory Syncytial Virus Disease (ERD) Made Simple," *Pediatric Research* 62 (2007): 111–115, https://doi.org/10.1203/PDR.0b013e3180686ce0.

8. "Synagis Prices and Coupons," WebMD, accessed August 28, 2021, https://www.webmd.com/rx/drug-prices/synagis.

9. "Synagis Helping Infants and Parents Breathe Easier: A Case Study," National Institutes of Health, October 23, 2002, https://www.ott.nih.gov/sites/default/files/documents/pdfs/SynagisCS.pdf.

10. James Watson, "How I Discovered DNA—James Watson," TED-Ed, July 26, 2013, video, https://www.youtube.com/watch?v=RvdxGDJogtA.

11. Steve Koppes, "Beam Us Up," Argonne National Library, January 3, 2019, https://www.anl.gov/article/beam-us-up.

12. Anthony Fauci, "Newscast," News4 This Week, December 15, 2013, video, @1:00, https://archive.org/details/WRC_20131215_103000_News4_This_Week/start/1260/end/1320.

CHAPTER 11: MERS

1. Islam Hussein, "The Story of the First MERS Patient," Nature Middle East, June 2, 2014, https://www.natureasia.com/en/nmiddleeast/article/10.1038/nmiddleeast.2014.134.

2. Jeremy Youde, "MERS and Global Health Governance," *International Journal* 70, no. 1 (2015): 119–136, https://doi.org/10.1177/0020702014562594.

3. "Middle East Respiratory Syndrome Coronavirus (MERS-CoV)," World Health Organization, accessed August 28, 2021, https://www.who.int/health-topics/middle-east-respiratory-syndrome-coronavirus-mers.

4. "Update: Severe Respiratory Illness Associated with Middle East Respiratory Syndrome Coronavirus (MERS-CoV)—Worldwide, 2012–2013," Centers for Disease Control and Prevention, June 7, 2013, https://www.cdc.gov/mmwr/preview/mmwrhtml/mm6223a6.htm.

5. Rui-Heng Xu et al., "Epidemiologic Clues to SARS Origin in China," *Emerging Infectious Diseases* 10, no. 6 (2004): 1031–1037, https://doi.org/10.3201/eid1006.030852.

6. "CDC SARS Response Timeline," Centers for Disease Control and Prevention, last modified April 26, 2013, https://www.cdc.gov/about/history/sars/timeline.htm.

7. "HIV Vaccine Study Halted by US Government over Unsuccessful Shots," *The Guardian*, April 25, 2013, https://www.theguardian.com/society/2013/apr/25/hiv-aids-vaccine-study-us-government.

8. Shan Shu et al., "Learning from the Past: Development of Safe and Effective COVID-19 Vaccines," *Nature Reviews Microbiology* 19 (2021): 211–219, https://doi.org/10.1038/s41579-020-00462-y.

9. Robert Kirchdoerfer et al., "Pre-fusion Structure of a Human Coronavirus Spike Protein," *Nature* 531 (2016): 118–121, https://doi.org/10.1038/nature17200.

CHAPTER 12: ZIKA

1. "The History of Zika Virus," World Health Organization, February 7, 2016, https://www.who.int/news-room/feature-stories/detail/the-history-of-zika-virus.

2. Gabriel Yan et al., "Distinguishing Zika and Dengue Viruses through Simple Clinical Assessment, Singapore," *Emerging Infectious Diseases* 24, no. 8 (2018): 1565–1568, https://doi.org/10.3201/eid2408.171883.

3. "Timeline of Emergence of Zika Virus in the Americas," Pan American Health Organization, accessed August 28, 2021, https://www3.paho.org/hq/index.php?option=com_content&view=article&id=11959:timeline-of-emergence-of-zika-virus-in-the-americas.

4. "The History of Zika Virus."

5. Gubio Campos et al., "Zika Virus Outbreak, Bahia, Brazil," *Emerging Infectious Diseases* 21, no. 10 (2015): 1885–1886, https://doi.org/10.3201/eid2110.150847.

6. Simon Romero, "Alarm Spreads in Brazil over a Virus and a Surge in Malformed Infants," *New York Times*, December 30, 2015, https://www.nytimes.com/2015/12/31/world/americas/alarm-spreads-in-brazil-over-a-virus-and-a-surge-in-malformed-infants.html.

7. "The History of Zika Virus."

8. "The History of Zika Virus."

9. "Timeline of Emergence of Zika Virus in the Americas."

10. "Zika Virus," Baylor College of Medicine, accessed August 28, 2021, https://www.bcm.edu/departments/molecular-virology-and-microbiology/emerging-infections-and-biodefense/specific-agents/zika.

11. "Spillover—Zika, Ebola & Beyond," PBS, August 2, 2016, video, https://www.pbs.org/video/spillover-zika-ebola-beyond-spillover-zika-ebola-beyond.

12. "Spillover—Zika, Ebola & Beyond."

13. "Zika Virus," World Health Organization, July 20, 2018, https://www.who.int/en/news-room/fact-sheets/detail/zika-virus.

14. Donald McNeil Jr., "Zika Virus, a Mosquito-Borne Infection, May Threaten Brazil's Newborns," *New York Times*, December 28, 2015, https://www.nytimes.com/2015/12/29/health/zika-virus-brazil-mosquito-brain-damage.html.

15. "Timeline of Emergence of Zika Virus in the Americas."

16. Ed Yong, "How Zika Conquered the Americas," *The Atlantic*, May 24, 2017, https://www.theatlantic.com/science/archive/2017/05/how-zika-conquered-the-americas/527961.

17. "Spillover—Zika, Ebola & Beyond."

18. "2014–2016 Ebola Outbreak in West Africa," Centers for Disease Control and Prevention, last modified March 8, 2019, https://www.cdc.gov/vhf/ebola/history/2014-2016-outbreak/index.html.

19. Sabrina Tavernise, "Zika Virus 'Spreading Explosively' in Americas, W.H.O. Says," *New York Times*, January 28, 2016, https://www.nytimes.com/2016/01/29/health/zika-virus-spreading-explosively-in-americas-who-says.html.

20. Mark Landler, "Obama Asks Congress for $1.8 Billion to Combat Zika Virus," *New York Times*, February 8, 2016, https://www.nytimes.com/2016/02/09/us/politics/obama-congress-funding-combat-zika-virus.html.

21. "Spillover—Zika, Ebola & Beyond."

22. Adrienne LaFrance, "Mosquitoes Are Spreading Zika in Florida," *The Atlantic*, July 29, 2016, https://www.theatlantic.com/health/archive/2016/07/zika-mosquitoes-officially-in-the-us/493610.

23. Adrienne LaFrance, "A Threat Bigger than Zika," *The Atlantic*, July 18, 2016, https://www.theatlantic.com/health/archive/2016/07/a-threat-bigger-than-zika/491564.

24. Anthony Fauci, "NIAID Director Dr. Anthony Fauci Addresses Zika Concerns," Fox News, August 22, 2016, video, https://www.youtube.com/watch?v=S2bPPLEtII4.

25. Barack Obama, "President Obama Receives a Briefing on the Zika Virus," Obama White House, May 20, 2016, video, https://www.youtube.com/watch?v=Sy13dYcxVXo.

26. "Safety and Immunogenicity of a Zika Virus DNA Vaccine, VRC-ZKADNA085-00-VP, in Healthy Adults," US National Library of Medicine, July 21, 2016, https://clinicaltrials.gov/ct2/show/NCT02840487.
27. "Zika Virus Vaccines," National Institute of Allergy and Infectious Diseases, last modified August 16, 2018, https://www.niaid.nih.gov/diseases-conditions/zika-vaccines.
28. "Zika Virus Vaccines."
29. Rick Bright, "BARDA-Supported Zika Vaccine Candidate Enters Clinical Trial," Public Health Emergency, January 9, 2017, https://www.phe.gov/ASPRBlog/Pages/BlogArticlePage.aspx?PostID=223.
30. Bright, "BARDA-Supported Zika Vaccine Candidate."
31. "VRC 705: A Zika Virus DNA Vaccine in Healthy Adults and Adolescents (DNA)," US National Library of Medicine, April 12, 2017, https://clinicaltrials.gov/ct2/show/NCT03110770.
32. Jon Cohen, "New Vaccine Coalition Aims to Ward Off Epidemics," Science, September 2, 2016, https://www.science.org/content/article/new-vaccine-coalition-aims-ward-epidemics-rev2.
33. Stanley Plotkin et al., "Establishing a Global Vaccine-Development Fund," *New England Journal of Medicine* 373, no. 4 (2015): 297–300, https://doi.org/10.1056/NEJMp1506820.
34. Jesper Pallesen et al., "Immunogenicity and Structures of a Rationally Designed Prefusion MERS-CoV Spike Antigen," *Proceedings of the National Academy of Sciences of the United States of America* 114, no. 35 (2017): E7348–E7357, https://doi.org/10.1073/pnas.1707304114.
35. Barney Graham et al., "Prefusion Coronavirus Spike Proteins and Their Use," United States patent 10,960,070, filed October 25, 2017, and March 30, 2021.
36. Melanie Swift, "The Johnson & Johnson Adenovirus Vaccine Explained," Mayo Clinic, April 30, 2021, https://www.mayoclinic.org/johnson-johnson-adenovirus-vaccine-explained/vid-20510091.
37. Maggie Fox, "Johnson & Johnson Booster Shot Prompts Large Increase in Immune Response, Company Says," CNN, August 25, 2021, https://www.cnn.com/2021/08/25/health/johnson-vaccine-booster-data/index.html

CHAPTER 13: THE RACE

1. Ja'Nel Johnson, "'Risk Is Extremely Low for Sacramento County' | First Case of Coronavirus Confirmed in Sacramento County," ABC10, February 21, 2020, https://www.abc10.com/article/news/local/sacramento/

sacramento-county-confirms-first-case-of-covid-19/103-551a22ff-2b3f-4153-b209-1ced7e13a577.

2. "Program Detail," Moderna, accessed August 31, 2021, https://investors.modernatx.com/program-detail.

3. "MRNA.OQ—Q4 2019 Moderna Inc Earnings Call," Thomson Reuters, February 26, 2020, https://investors.modernatx.com/static-files/c78d1ffc-cdb1-4b4c-bd6d-e50d96ca3cf0.

4. Robert Costa and Philip Rucker, "Woodward Book: Trump Says He Knew Coronavirus Was 'Deadly' and Worse than the Flu while Intentionally Misleading Americans," *Washington Post*, September 9, 2020, https://www.washingtonpost.com/politics/bob-woodward-rage-book-trump/2020/09/09/0368fe3c-efd2-11ea-b4bc-3a2098fc73d4_story.html.

5. Rob Stein, "NIH Lab Races to Create Coronavirus Vaccine in Record Time," NPR, February 21, 2020, https://www.npr.org/2020/02/21/808016544/nih-lab-races-to-create-coronavirus-vaccine-in-record-time.

6. Will Feuer and Kevin Stankiewicz, "CDC to Test More Suspected Cases of Coronavirus after Revising Guidelines," CNBC, February 27, 2020, https://www.cnbc.com/2020/02/27/cdc-to-test-more-suspected-cases-of-coronavirus-after-revising-guidelines.html.

7. Matthew Herper, "He Experienced a Severe Reaction to Moderna's Covid-19 Vaccine Candidate. He's Still a Believer," *Stat*, May 26, 2020, https://www.statnews.com/2020/05/26/moderna-vaccine-candidate-trial-participant-severe-reaction.

8. Barney Graham, "Rapid COVID-19 Vaccine Development," *Science* 368, no. 6494 (2020): 945–946, https://doi.org/10.1126/science.abb8923.

9. Sanjay Gupta, *Race for the Vaccine*, film, CNN, 2021.

10. "Moderna Loses Key Patent Challenge," *Nature Biotechnology,* September 4, 2020, https://www.nature.com/articles/s41587-020-0674-1.

11. "Statement from Moderna on Patent Trial and Appeal Board (PTAB) Ruling," Moderna, July 24, 2020, https://investors.modernatx.com/news-releases/news-release-details/statement-moderna-patent-trial-and-appeal-board-ptab-ruling.

EPILOGUE

1. Abe Aboraya, "FHA Survey: 68 Florida Hospitals Have Less than 48 Hours Worth of Oxygen," WUSF, August 26, 2021, https://wusfnews.wusf.usf.edu/health-news-florida/2021-08-26/survey-68-florida-hospitals-have-less-than-48-hours-worth-of-oxygen.

2. Dave Muoio, "10 States Nearing—or Exceeding—Hospital Capacity during COVID's Summer Resurgence," Fierce Healthcare, August 19, 2021, https://www.fiercehealthcare.com/hospitals/10-states-nearing-or-exceeding-hospital-capacity-during-covid-s-summer-resurgence.

3. Jay Bhattacharya and Donald Boudreaux, "Eradication of Covid Is a Dangerous and Expensive Fantasy," *Wall Street Journal*, August 4, 2021, https://www.wsj.com/articles/zero-covid-coronavirus-pandemic-lockdowns-china-australia-new-zealand-11628101945.

4. "COVID-19 Vaccinations in the United States," Centers for Disease Control and Prevention, accessed August 31, 2021, https://covid.cdc.gov/covid-data-tracker/#vaccinations_vacc-total-admin-rate-total.

5. Michaeleen Doucleff, "The Delta Variant Isn't as Contagious as Chickenpox. But It's Still Highly Contagious," *NPR*, August 11, 2021, https://www.npr.org/sections/goatsandsoda/2021/08/11/1026190062/covid-delta-variant-transmission-cdc-chickenpox.

6. "State School Immunization Requirements and Vaccine Exemption Laws," Centers for Disease Control and Prevention, accessed September 1, 2021, https://www.cdc.gov/phlp/docs/school-vaccinations.pdf.

7. "Transmission of Measles," Centers for Disease Control and Prevention, last modified November 5, 2020, https://www.cdc.gov/measles/transmission.html.

8. Carrie Macmillan, "Herd Immunity: Will We Ever Get There?" Yale Medicine, May 3, 2021, https://www.yalemedicine.org/news/herd-immunity.

9. Elena Conis, "Measles and the Modern History of Vaccination," *Public Health Reports* 134, no. 2 (2019): 118–125, https://doi.org/10.1177/0033354919826558.

10. "Measles Cases and Outbreaks," Centers for Disease Control and Prevention, last modified July 9, 2021, https://www.cdc.gov/measles/cases-outbreaks.html.

11. "Measles History," Centers for Disease Control and Prevention, last modified November 5, 2020, https://www.cdc.gov/measles/about/history.html.

12. Dan Diamond, "More than 150 Employees Resign or Are Fired from Houston Hospital System after Refusing to Get Vaccinated," *Texas Tribune*, June 23, 2021, https://www.texastribune.org/2021/06/23/texas-hospital-houston-methodist-vaccine-employees-fired-resign.

13. *Bridges et al. vs Houston Methodist Hospital et al.*, United States District Court, Southern District of Texas, Civil Action H-21-1774, Order of Dismissal, June 12, 2021.

14. Wendy Mariner et al., "*Jacobson v Massachusetts*: It's Not Your Great-Great-Grandfather's Public Health Law," *American Journal of Public Health* 95, no. 4 (2005): 581–590, https://doi.org/10.2105/AJPH.2004.055160.

15. MSNBC (@MSNBC), "Mehdi Hasan: 'With the vaccine reluctant or hesitant . . . can we get to herd immunity without them?' Dr. Fauci: 'I don't think so,'" Twitter, August 24, 2021, https://twitter.com/MSNBC/status/1430134039734718469.

16. "Public Trust in CDC, FDA, and Fauci Holds Steady, Survey Shows," Annenberg Public Policy Center, July 20, 2021, https://www.annenbergpublicpolicycenter.org/public-trust-in-cdc-fda-and-fauci-holds-steady-survey-shows.

17. "Trump Rally Crowd Chants 'Fire Fauci,'" CBS News, November 2, 2020, video, https://www.youtube.com/watch?v=sECFxmDJ6aE.

18. "Trump Wants You to Believe Coronavirus Cases Are 'Up Because We TEST.' He's Wrong,' Vox, October 26, 2020, https://www.vox.com/2020/10/26/21534380/trump-coronavirus-cases-spike-testing.

19. Victoria Knight and Julie Appleby, "How COVID Death Counts Become the Stuff of Conspiracy Theories," *Kaiser Health News*, November 2, 2020, https://khn.org/news/how-covid-death-counts-become-the-stuff-of-conspiracy-theories.

20. Jordan Mendoza, "Donald Trump Booed at Alabama Rally after Telling Supporters to 'Take the Vaccines,'" *USA Today*, August 23, 2021, https://www.usatoday.com/story/news/politics/2021/08/23/donald-trump-tells-alabama-rally-covid-19-vaccine-gets-booed/8237487002.

21. JoNel Aleccia, "These Guvs Push Antibodies—but Oppose Vax and Mask Mandates," *Daily Beast*, August 25, 2021, https://www.thedailybeast.com/govs-ron-desantis-and-greg-abbott-push-regeneron-monoclonal-antibodies-oppose-mandates.

22. Hannah Rodriguez, "Some Iowa Farm Store Customers Seeking Unauthorized Animal Deworming Drug Ivermectin for COVID-19 Treatment," *Des Moines Register*, August 26, 2021, https://www.desmoinesregister.com/story/money/business/retail/2021/08/26/some-iowa-farm-store-customers-seeking-animal-deworming-drug-ivermectin-covid-treatment-fda-warning/5604155001.

23. "President Honours Dr Salk Aka Dr Salk Honoured (1955)," British Pathé, April 13, 2014, video, https://www.youtube.com/watch?v=kIj92iKwvjI.

24. "Press Briefing by White House COVID-19 Response Team and Public Health Officials," White House, April 9, 2021, https://www.whitehouse.gov/briefing-room/press-briefings/2021/04/09/press-briefing-by-white-house-covid-19-response-team-and-public-health-officials-26.

25. Bill Gates, "The Next Outbreak? We're Not Ready," TED, March 2015, video, @8:24, https://www.ted.com/talks/bill_gates_the_next_outbreak_we_re_not_ready#t-20580.

26. Bill Gates, "What Bill Gates Is Afraid Of," Vox, May 27, 2015, video, https://www.youtube.com/watch?v=9AEMKudv5p0.

27. Berkeley Lovelace, "Pfizer CEO Sees a Return to 'Normal' Globally by the End of 2022," CNBC, June 16, 2021, https://www.cnbc.com/2021/06/16/covid-pfizer-ceo-sees-a-return-to-normal-globally-at-the-end-of-2022.html.

28. Benjamin Mueller and Rebecca Robbins, "Where a Vast Global Vaccination Program Went Wrong," *New York Times*, August 2, 2021, https://www.nytimes.com/2021/08/02/world/europe/covax-covid-vaccine-problems-africa.html.

29. Jennifer Griffin et al., "SARS-CoV-2 Infections and Hospitalizations Among Persons Aged ≥16 Years, by Vaccination Status—Los Angeles County, California, May 1–July 25, 2021," *Morbidity and Mortality Weekly Report* 70, no. 34 (2021): 1170–1176, http://doi.org/10.15585/mmwr.mm7034e5.

30. Ashley Fowlkes et al., "Effectiveness of COVID-19 Vaccines in Preventing SARS-CoV-2 Infection Among Frontline Workers Before and During B.1.617.2 (Delta) Variant Predominance—Eight U.S. Locations, December 2020–August 2021," *Morbidity and Mortality Weekly Report* 70, no. 34 (2021): 1167–1169, http://doi.org/10.15585/mmwr.mm7034e4.

31. Berkeley Lovelace, "Israel Says Pfizer Covid Vaccine Is Just 39% Effective as Delta Spreads, but Still Prevents Severe Illness," CNBC, July 23, 2021, https://www.cnbc.com/2021/07/23/delta-variant-pfizer-covid-vaccine-39percent-effective-in-israel-prevents-severe-illness.html.

ABOUT THE AUTHOR

David Heath is an award-winning investigative journalist. He has worked at several news organizations, including CNN, *USA Today*, and the Center for Public Integrity. His work has helped to change policies and laws and has even led to criminal indictments. His articles have exposed illegal insider trading in drug research, questionable medical research and political influence to derail tougher toxic-chemical standards. His work has appeared on PBS Frontline, PBS NewsHour, the CBS Evening News, the Huffington Post, *The Atlantic*, *Time*, *Newsweek*, the Daily Beast, Vice News, *Scientific American*, and *Mother Jones*. He has won more than two dozen national journalism awards, including the Goldsmith, the Gerald Loeb, and George Polk. His work with others has been nominated for a national Emmy and he has been a finalist for the Pulitzer Prize three times. He lives in Bethesda, Maryland.